ADVANCES IN HYDROGEN ENERGY

ADVANCES IN HYDROGEN ENERGY

Edited by

Catherine E. Grégoire Padró
National Renewable Energy Laboratory
Golden, Colorado

and

Francis Lau
Institute of Gas Technology
Des Plaines, Illinois

Kluwer Academic / Plenum Publishers
New York, Boston, Dordrecht, London, Moscow

Proceedings of an American Chemical Society Symposium on Hydrogen Production, Storage, and Utilization, held August 22–26, 1999, in New Orleans, Louisiana

ISBN: 0-306-46429-2

©2000 Kluwer Academic / Plenum Publishers, New York
233 Spring Street, New York, N.Y. 10013

http://www.wkap.nl/

10 9 8 7 6 5 4 3 2 1

A C.I.P. record for this book is available from the Library of Congress

Printed in the United States of America

FOREWORD

In the future, our energy systems will need to be renewable and sustainable, efficient and cost-effective, convenient and safe. Hydrogen has been proposed as the perfect fuel for this future energy system. The availability of a reliable and cost-effective supply, safe and efficient storage, and convenient end use of hydrogen will be essential for a transition to a Hydrogen Economy. Research is being conducted throughout the world for the development of safe, cost-effective hydrogen production, storage, and use technologies that support and foster this transition. This book is a collection of important research and analysis papers on hydrogen production, storage, and end use technologies that were presented at the American Chemical Society National Meeting, in New Orleans, Louisiana (USA), in August, 1999.

Hydrogen production from fossil fuels will continue for the foreseeable future, given the large resource and the established industrial base. Research is focused on improving the environmental aspects of fossil fuel use, and a number of papers address advanced hydrogen production technologies that reduce or eliminate CO_2 emissions from the production process. In addition, hydrogen production from biomass, a renewable resource with the potential for zero net CO_2 emissions, is discussed.

Hydrogen production technologies, no matter the feedstock, rely on hydrogen separation and purification technologies that, due to high energy consumption, reduce the overall efficiency of the process. The development of membrane separation technologies that reduce the amount of energy required to produce high-purity hydrogen will have important impacts on current and future production processes. Ceramic membrane development work is discussed in this book.

Understanding the behavior of hydrogen in metals and alloys is important for the development of efficient hydrogen storage and transport processes. Several papers focus on developing a fundamental understanding of the effect of hydrogen on metals. In addition, a novel storage and transport process relying on chemical hydride slurries is also presented.

Safety aspects of hydrogen use are of primary concern, particularly given the negative public perception of hydrogen. Papers cover the development of cost-effective hydrogen sensors for use in vehicles and buildings, and discuss insights gained in the design of self-venting buildings via modeling of the behavior of hydrogen leaks in closed spaces.

Finally, the development of integrated hydrogen energy systems is discussed from the perspective of grid-independent renewable systems for remote applications.

We believe the papers in this book will serve to advance the concept of a Hydrogen Economy as a technical, economic, and environmental solution to increased energy consumption and a cleaner world.

Catherine E. Grégoire Padró
National Renewable Energy Laboratory
and
Francis Lau
Institute of Gas Technology

CONTENTS

HYDROGEN FROM FOSSIL FUELS WITHOUT CO_2 EMISSIONS

Nazim Muradov

Florida Solar Energy Center
Cocoa, FL 32922-5703

INTRODUCTION

In the near- to medium-term future, hydrogen production will continue to rely on fossil fuels, primarily natural gas (NG). It is generally understood that the renewable energy-based processes of hydrogen production (photoelectrochemical and photobiological decomposition of water, solar-photovoltaic water electrolysis, thermochemical and hybrid water splitting cycles, etc.) would unlikely yield significant reduction in hydrogen costs in the next 1-2 decades. The future of nuclear power systems, a relatively clean and abundant energy source, still remains uncertain due to strong public opposition. In general, given the advantages inherent in fossil fuels such as their availability, cost-competitiveness, convenience of storage and transportation, they are likely to play a major role in global energy supply for the next century. On the other hand, fossil fuels are major source of anthropogenic CO_2 emissions to the atmosphere. Various scenarios of global energy use in the next century predict a continued increase in CO_2 emissions that would gradually rise its concentration in the atmosphere to dangerous levels. It is clear that the industrialized world would not be able to retain present living standards and meet challenges of global warming, unless major changes are made in the way we produce energy, and manage carbon emissions.

There are several possible ways to mitigate CO_2 emission problems. Among them are traditional approaches including: (i) more efficient use of fossil fuel energy resources, (ii) increased use of clean fossil fuels, such as NG, and (iii) increased use of non-fossil fuels (nuclear power and renewable sources). The novel and most radical approach to effectively manage carbon emissions is the decarbonization of fossil fuels. Three main scenarios of fossil fuels decarbonization are currently discussed in the literature:

- CO_2 sequestration after fossil fuel combustion in energy conversion devices
- production of hydrogen by conventional processes (steam reforming, partial oxidation, etc.) with subsequent CO_2 sequestration
- production of hydrogen and carbon via decomposition of NG and hydrocarbon fuels

Advances in Hydrogen Energy, edited by Padró and Lau
Kluwer Academic/Plenum Publishers, 2000

The objective of this paper is to discuss different strategies of decarbonization of fossil fuels (mainly, NG) via hydrogen production (second and third scenario).

Commercial Hydrogen Production Processes

Steam reforming. Steam reforming (SR) of NG is the most efficient and widely used process for the production of hydrogen. The SR process basically represents catalytic conversion of methane (a major component of the hydrocarbon feedstock) and water (steam) to hydrogen and carbon oxides, and consists of two main reactions:

Synthesis gas generation: $\qquad$ $CH_4 + H_2O \rightarrow 3H_2 + CO$

Water-gas shift (WGS) reaction: $\qquad$ $CO + H_2O \rightarrow H_2 + CO_2$

Overall reactiton: $\qquad$ $CH_4 + 2H_2O \rightarrow 4H_2 + CO_2 + 163 \text{ kJ}$

Four moles of hydrogen are produced in the process, with half coming from the methane and half from water. The theoretical energy requirement per mole of hydrogen produced for the overall process is equal to $163:4 = 40.75$ kJ/mol H_2. The SR process, in general, is very energy intensive since it operates at high temperatures (850-950°C) and pressure (35 atm). The process is favored by high steam/carbon ratios (3-5), and its fuel usage is a significant part (30-40%) of the total NG usage of a typical hydrogen plant. This significantly reduces overall thermal efficiency of the process. Thus, the SR process thermal efficiency is seldom greater than 50%.[1] There is no by-product credit (except for steam) for the process and, in the final analysis, it does not look environmentally benign due to large CO_2 emissions. The total CO_2 emissions (including stack gases) from SR process reach up to 0.3-0.4 m³ CO_2 per each m³ of hydrogen produced.

Partial oxidation. In partial oxidation (PO) processes, a fuel and oxygen (and sometimes steam) are combined in proportions such that the fuel is converted into a mixture of H_2 and CO. There are several modifications of PO process, depending on the composition of the process feed and type of the fossil fuel used. The overall process is exothermic, due to a sufficient amount of oxygen that is added to the reagent stream. PO processes can be carried out catalytically or non-catalytically. The non-catalytic PO process operates at high temperatures (1100-1500°C), and it can use any possible feedstock, including heavy residual oils and coal. The catalytic process is carried out at significantly lower range of temperatures (600-900°C) and, in general, uses light hydrocarbon fuels as feedstocks, e.g. NG and naphtha. PO of methane can be described by the following equations:

$CH_4 + 1/2O_2 \rightarrow CO + 2H_2$ $\qquad$ $\Delta H° = -35.6$ kJ/mol

$CH_4 + O_2 \rightarrow CO_2 + 2H_2$ $\qquad$ $\Delta H° = -319.3$ kJ/mol

Two moles of hydrogen are produced for every mole of methane. Both reactions are exothermic, which implies that the reactor does not need an external heat source. If pure oxygen is used in the process, it has to be produced and stored, significantly adding to the cost of the system. On the other hand, if PO process uses air as an oxidizer, the effluent gas would be heavily diluted by nitrogen resulting in larger WGS reactor and gas purification units. The maximum theoretical concentration of hydrogen in the effluent gas using pure oxygen is 66.7

% by volume, however, the concentration drops to 40.9 % if air is used as an oxidizer. The amount of CO_2 produced by PO process depends on the composition of the feedstock used.

Steam-Iron Process. The steam-iron (SI) process has long been practiced for the production of hydrogen from a wide variety of fossil fuels, including coal. The SI process produces high-purity hydrogen by separating the hydrogen production and fuel oxidation steps using an iron oxide reduction-oxidation regenerative system. Thus, it does not require WGS and CO_2 removal stages. Recently, the SI process was modified for fuel cell applications. The sponge iron is oxidized in a multiple bed reactor to provide high-purity hydrogen to a fuel cell, while already depleted beds are regenerated (reduced) using synthesis gas delivered from a methane-fueled steam reformer. However, the process is multi-stage, requires high temperatures (for the reduction of the magnetite, Fe_3O_4, to sponge iron) and additional step of NG steam reforming. Practically all carbon present in a hydrocarbon fuel used for hydrogen production is finally transformed into CO_2 and vented to the atmosphere.

Thermal Decomposition (TD). The objective of TD process is to thermally decompose hydrocarbon fuel, particularly NG, into its constituent elements: hydrogen and carbon. Originally, TD has been employed for the production of carbon black with hydrogen being a byproduct and supplementary fuel for the process. TD has been practiced in a semi-continuous mode using two tandem reactors at high temperatures. After one of the reactors was heated to the pyrolysis temperature of approximately 1400°C by a fuel-air flame, the air was cut off. The hydrocarbon was pyrolyzed over the heated contact (firebrick) into hydrogen and carbon black particles. Simultaneously, another reactor was heated to pyrolysis temperature, followed by the reversing flow of hydrocarbon feedstock from the pyrolysis reactor to the heated reactor, and the process continued in a cyclic mode. Carbon black produced by this process has been mainly used in the tire and pigment industries. Currently, TD processes, as a source of carbon black, have very limited application, supplanted by more efficient, continuous furnace black processes based on partial oxidation of petroleum feedstock.

CO$_2$ SEQUESTRATION

The perspectives of CO_2 sequestration is actively discussed in the literature.[2-4] The main objective of carbon sequestration is to prevent anthropogenic CO_2 emissions from reaching the atmosphere by capturing and securely storing CO_2 underground or in the ocean. Of particular interest is the sequestration of CO_2 produced by conventional hydrogen production processes (e.g. SR, PO). If these hydrogen production technologies could be coupled with CO_2 sequestration, there would be practically no environmental constraints on using fossil fuels on a large scale. A typical hydrogen plant with the capacity of approximately one million m^3 of hydrogen per day produces about 0.25 million standard cubic meters of CO_2 per day (exclusive of stack gases), which is normally vented into the atmosphere. CO_2 concentrations in process streams range from approximately 5 vol% for stack gases to almost 100 vol% for concentrated streams from pressure swing adsorption (PSA) (or other advanced gas separation systems). The present-day options for CO_2 capture and separation include:

- absorption (chemical and physical)
- adsorption (chemical and physical)
- low-temperature distillation
- membrane separation.

There have been estimates reported in the literature on the economics of CO_2 sequestration associated with hydrogen production from fossil fuels. The capture and disposal of CO_2 (80-85% of CO_2 captured from the concentrated streams of SR process) add about 25-30% to the cost of hydrogen produced by the SR of NG.[4] Since all stages of the CO_2 sequestration process, including its capture, pressurization, transportation and injection underground (or in the ocean), are energy intensive processes, it was important to estimate the total energy consumption per unit of CO_2 sequestrated. The following summarizes available data on the energy consumption during CO_2 sequestration associated with the SR process (per kg of sequestrated CO_2):

- CO_2 capture by hot K_2CO_3 solutions, ~3000 kJ[5]
- CO_2 pressurization to 80 bar by 5 stage compression, 281 kJ_{el}[3]
- CO_2 pipeline transportation for 100-500 km to the disposal site and injection, ~2000 kJ_{el}

About 80% of the world's commercial energy is based on fossil fuels[2] (84% for U.S.A [6]). World average for CO_2 emission associated with electricity production is 0.153 kg of CO_2 per kWh produced.[3] Thus, the total CO_2 emissions from CO_2 sequestration are estimated at 0.20-0.25 kg CO_2 per kg of sequestrated CO_2. Because CO_2 sequestration is an energy intensive process, in the final analysis, it does not completely eliminate CO_2 emission. In addition to this problem, some uncertainties remain regarding the duration and extent of CO_2 retention (underground or in the ocean) and its possible environmental effect.

ADVANCED PROCESSES OF METHANE DECOMPOSITION

Methane is one of the most stable organic molecules. Its electronic structure, lack of polarity and any functional group makes it extremely difficult to decompose into its constituent elements. Several novel approaches to the problem of methane decomposition into hydrogen and carbon, and valuable hydrocarbons are discussed in this section.

Thermal Systems

Methane decomposition reaction is a moderately endothermic process:

$$CH_4 \rightarrow C + 2H_2 \qquad \qquad \Delta H° = 75.6 \text{ kJ/mol}$$

The energy requirement per mole of hydrogen produced (37.8 kJ/mol H_2) is somewhat less than for the SR process. Due to a relatively low endothermicity of the process, less than 10% of the heat of methane combustion is needed to drive the process. In addition to hydrogen as a major product, the process produces a very important byproduct: clean carbon. The process does not include the WGS reaction and energy-intensive gas separation stages. A preliminary process design for a continuous methane decomposition process and its economics has been developed.[7] The techno-economic assessment showed that the cost of hydrogen produced by TD of NG ($58/1000 m³ H_2, with carbon credit), was somewhat lower than that for the SR process ($67/1000 m³ H_2).[7]

Recently, several new processes for methane thermal decomposition were reported in the literature. In one report, the authors proposed a methane decomposition reactor consisting of a molten metal bath.[8] Methane bubbles through molten tin or copper bath at high temperatures (900°C and higher). The advantages of this system are: efficient heat transfer to a methane gas stream and ease of carbon separation from the liquid metal surface by density difference. In another work, methane decomposition was carried out in a continuous process using a metallic tubular reactor in the range of temperatures 700-900°C and pressures 28.2-56.1 atm.[9] It was shown that at 900°C, 56.1 atm and sufficiently high residence time (>100 sec) the concentration of methane in the effluent gas approached equilibrium conditions. The determined reaction activation energy of E_a= 131.1 kJ/mol was substantially lower than E_a reported in the literature for homogeneous methane decomposition (272.4 kJ/mol), pointing to a significant contribution of the heterogeneous processes caused by the submicron size carbon particles adhered to the reactor surface. Finally, a high temperature regenerative gas heater (HTRGH) for hydrogen and carbon production from NG has been developed.[10] In this process thermal decomposition of NG was conducted in the "free volume" of HTRGH using carrier gas (N_2 or H_2) pre-heated up to 1627-1727°C in the matrix of the regenerative gas heater. The reactor was combined with a steam turbine to increase the overall efficiency of the system.

Fast Pyrolysis of Methane in Tubular Reactor. We have conducted a series of experiments on fast pyrolysis of methane using ceramic (alumina) and quartz tubular reactors. The objective was to thermally (homogeneously) decompose methane to hydrogen, carbon and valuable unsaturated and aromatic hydrocarbons. Preliminary testing of the catalytic activity of quartz and alumina toward methane decomposition proved their inertness at temperatures below 1100°C. Tubular reactors with internal diameters of 3-6 mm and a small reaction zone with residence times in the range of 1-20 milliseconds were used in these experiments. Preheated (400°C) methane streams entered the reactor at flow rates in the range of 1-10 liters/min and were subjected to pyrolysis at the temperatures of 900-1100°C. The conversion of methane was found to be a function of the temperature and residence time. For example, at the reaction zone temperature of 1100°C and residence times of 1.0, 2.0 and 6.2 ms, methane conversions were 0.1, 2.0 and 16.1%, respectively. Hydrogen and carbon were the main products of pyrolysis accounting for more than 80 wt% of the products. Unsaturated (mostly, C_2) and aromatic (including polynuclear) hydrocarbons were also produced in significant quantities as byproducts of methane pyrolysis. For example, at the reaction zone temperature of 1100°C and the residence time of 6.2 ms, the yields of gaseous and liquid products were as follows (mol%): C_2H_6- 0.9, C_2H_4- 3.3, C_2H_2- 5.8, C_2-C_6- 1.5, polynuclear aromatics (naphthalene, anthracene)- 2.0. Unidentified liquid products of pyrolysis accounted for approximately 5 wt% of methane pyrolysis products. Carbon (coke) was mostly deposited on the reactor wall down-stream of the reaction zone, which indicated that methane decomposition reaction occurred predominantly in gas phase. At higher residence times (seconds and minutes scale), the yields of C_2^+ and polyaromatic hydrocarbons dramatically dropped. These experiments demonstrated that the methane decomposition process could be arranged in a homogeneous mode producing not only hydrogen and carbon, but also a variety of very valuable hydrocarbons (ethylene, acetylene, aromatics).

The mechanism of thermal decomposition (pyrolysis) of methane has been extensively studied.[11] Since C- H bonds in methane molecule are significantly stronger than C-H and C-C bonds of the products, secondary and tertiary reactions contribute at the very early stages of the reaction, obscuring the initial processes. It has been shown[11] that the homogeneous dissociation of methane is the only primary source of free radicals and controls the rate of the overall process:

$$CH_4 \rightarrow CH_3^{\cdot} + H^{\cdot}$$

This reaction is followed by a series of consecutive and parallel reactions with much lower activation energies. After the formation of acetylene (C_2H_2), a sequence of very fast reactions occurs leading to the production of higher unsaturated and aromatic hydrocarbons and finally carbon:

$$C_2H_2 \rightarrow \text{high unsaturated hydrocarbons} \rightarrow \text{aromatics} \rightarrow \text{polynuclear aromatics} \rightarrow \text{carbon}$$

This involves simultaneous decomposition and polymerization processes and phase changes from gas to liquid to solid. A detailed mechanism of the final transformation to carbon is rather complex and is not well understood.

Plasma Decomposition

Plasma-assisted decomposition of hydrocarbons with the production of hydrogen and carbon has become an active area of research recently. Kvaerner company of Norway has developed a methane decomposition process that produces hydrogen and carbon black by using high temperature plasma (CB&H process).[12] The advantages of thermal plasma process are: high thermal efficiency (>90%), high fuel flexibility, purity of hydrogen (98 vol%) and production of valuable byproduct- carbon. Very low CO_2 emissions are associated with the plasma process.

In another paper, the authors advocated a plasma-assisted decomposition of methane into hydrogen and carbon.[13] It was estimated that 1-1.9 kWh of electrical energy is consumed per normal cubic meter of hydrogen produced. The authors stated that plasma production of hydrogen is free of CO_2 emissions. However, since most of the electric energy supply in the world comes from fossil fuels, electricity-driven hydrogen production processes such as plasma and electrochemical processes, are CO_2 producers.

Photolysis

Due to the high dissociation energy of CH_3 - H bond (D_o= 4.48 eV), methane absorbs irradiation in the vacuum ultra-violet region. The absorption spectrum of methane is continuous in the region from 1100 to 1600Å (absorption coefficient k= 500 $atm^{-1}cm^{-1}$ at 1100-1300Å). Unfortunately, wavelengths shorter than 1600 Å are present neither in the solar spectrum, nor in the output of most UV lamps. Therefore, production of hydrogen and other products by direct photolysis of methane does not seem practical.

Photocatalytic Activation of Methane. The methane molecule could be activated in the presence of special photocatalysts using near-UV photons that are present in the solar spectrum (up to 5% of total energy). Previously we have demonstrated photocatalytic conversion of low alkanes (C_1-C_3) to unsaturated hydrocarbons (mostly C_2-C_4 olefins) under UV irradiation using polyoxometalates of W, Mo, V and Cr.[14] The diffuse reflectance UV-VIS spectra of the synthesized silica-supported polyoxotungstates (POT, $[H_xW_yO_z]$) exhibit continuous absorption up to 350 nm (near-UV area). Irradiation of methane adsorbed to the surface of POT/SiO_2 with near-UV photons at room temperature resulted in the photoreduction of POT to its reduced (blue-colored) form with simultaneous photoconversion of methane to C_2^+ products. Thermal desorption (125-200°C) of products in vacuum resulted in the following gaseous mixture (vol%): C_2H_4- 40.2, C_3H_6- 21.7, C_4H_8- 36.0, C_5^+- 2.1. CO and CO_2 were not detected among the products of methane photo-transformation. The total products yield (per adsorbed methane) was 17%. However, doping the POT catalyst with Pt (0.1 wt%) increased the products yield to 32.1%. In the presence of UV-illuminated SiO_2-supported (5 wt%) silica-tungstic acid ($H_4SiW_{12}O_{40}$) 19.3% of the adsorbed methane was converted to the following gaseous mixture (vol%): C_2H_4- 4.1, C_3H_6- 7.3, C_4H_8- 86.2, C_5^+- 2.4.

The mechanism of methane photoactivation involves the initial photoinduced charge transfer in photoactive $W^{6+}=O$ groups of POT and STA molecules, leading to the formation of very active electron-deficient species able to abstract H-atom from a methane molecule:

$$W^{6+}=O + h\nu \rightarrow W^{5+}\text{-}O^{\cdot}$$

$$W^{5+}\text{-}O^{\cdot} + CH_4 \rightarrow W^{5+}\text{-}OH + CH_3^{\cdot}$$

$$W^{6+}=O + CH_3^{\cdot} \rightarrow W^{5+}\text{-}OCH_3$$

$$2W^{5+}\text{-}OCH_3 \rightarrow 2W^{5+}\text{-}OH + (C_2H_4)_{chem}$$

Higher molecular weight olefins (C_3H_6, C_4H_8, etc.) are, most likely, produced by the secondary catalyzed reactions of chemisorbed ethylene, $(C_2H_4)_{chem}$. After the photoreaction, the photocatalyst remains in its photoreduced form ($W^{5+}\text{-}OH$) at ambient temperature. It could be thermally regenerated to its initial oxidized form ($W^{6+}=O$) by releasing hydrogen:

$$2W^{5+}\text{-}OH \rightarrow 2W^{6+}=O + H_2$$

Thus, the overall reaction represents the photocatalytic transformation of methane to hydrogen and ethylene:

$$2CH_4 + h\nu \rightarrow C_2H_4 + 2H_2$$

Concentrated solar irradiation could be used to drive the thermal stages (desorption of products and regeneration of the catalyst) of this process. The advantage of this potentially solar-driven process is that it converts methane to hydrogen and valuable olefins without production of CO_2.

Thermocatalytic Decomposition

There have been attempts to use catalysts in order to reduce the maximum temperature of thermal decomposition of methane. In the 1960s, Universal Oil Products Co. developed the HYPRO[⫿] process for continuous production of hydrogen by catalytic decomposition of a gaseous hydrocarbon streams.[15] Methane decomposition was carried out in a fluidized bed catalytic reactor from 815 to 1093°C. Supported Ni, Fe and Co catalysts (preferably Ni/Al$_2$O$_3$) were used in the process. The coked catalyst was continuously removed from the reactor to the regeneration section where carbon was burned off by air, and the regenerated catalyst returned to the reactor. Unfortunately, the system with two fluidized beds and the solids-circulation system was too complex and expensive and could not compete with the SR process.

NASA has conducted studies on the development of catalysts for a methane decomposition process for space life support systems.[16] A special catalytic reactor with a rotating magnetic field to support Co-catalyst at 850°C was designed. In the 1970s, a group of U.S. Army researchers developed a fuel processor (conditioner) to catalytically convert different hydrocarbon fuels to hydrogen, used to feed a 1.5 kW fuel cell.[17] A stream of gaseous fuel entered one of two reactor beds, where hydrocarbon decomposition to hydrogen took place at 870-980°C and carbon was deposited on the Ni-catalyst. Simultaneously, air entered the second reactor where catalyst regeneration by burning coke off of the catalyst surface occurred. The streams of fuel and air to the reactors then were reversed for another cycle of decomposition-regeneration. The fuel processor did not require WGS and gas separation stages, a significant advantage. However, the thermal efficiency of this type of processor, in general, is relatively low (<60%) and they produce CO_2 in quantities comparable with SR and PO processes. Recently, several groups of researchers have reported on the development of hydrocarbon fuel processors for fuel cell applications using a similar concept.[18,19]

It was found that almost all transition metals (d-metals) exhibit catalytic activity toward methane decomposition reaction to some extent, and some demonstrate remarkably high activity. It should be noted, however, that there is no universal agreement among different groups of researchers regarding the choice of the most efficient metal catalyst for methane decomposition. For example, it was demonstrated that the rate of methane activation in the presence of transition metals followed the order: Co, Ru, Ni, Rh > Pt, Re, Ir > Pd, Cu, W, Fe, Mo.[20] Other researchers have found Pd to be the most active catalyst for methane decomposition,[18,21] whereas still others found Ni was the catalyst of choice,[22] or Fe and Ni.[23,24] Finally, Co catalyst demonstrated highest activity in methane decomposition reaction.[25]

Of particular interest are catalytic methane decomposition reactions producing special (e.g. filamentous) forms of carbon. For example, researchers have reported catalytic decomposition of methane over Ni catalyst at 500°C with the production of hydrogen and whisker carbon,[26] and concentrated solar radiation was used to thermally decompose methane into hydrogen and filamentous carbon.[27] The advantages of this system included efficient heat transfer due to direct irradiation of the catalyst, and CO_2-free operation.

Metal *vs* Carbon Catalysts

We have determined the catalytic activity of the wide range of metal catalysts and found that Ni/alumina and Fe/alumina catalysts exhibited very high initial activity in the methane decomposition reaction. For example, in the presence of freshly reduced Ni- catalysts, hydrogen was detected in the effluent gas at the temperature as low as 200°C. Figure 1 depicts the kinetic curve of hydrogen production over reduced Fe/Al_2O_3 catalyst at 850°C. The maximum hydrogen production yield was observed at the onset of the process, followed by the gradual decrease in hydrogen production rate and, finally, by steady-state decomposition of methane. A gradual decline in the hydrogen production rate could be attributed to carbon build-up on the catalyst surface. The shape of the kinetic curve is typical for methane decomposition in the presence of other transition metal catalysts.

Similar observations were reported by other researchers. For example, it was demonstrated that the values for the rate constants for methane decomposition over Ni, Co and Fe catalysts declined as the run proceeded.[28] At the temperatures below 1000°C the experimental data followed the kinetic equation:

$$-d[CH_4]/dt = k \, S \, (1-\theta)[CH_4]$$

where k is intrinsic rate constant, S is the surface area, and θ is the fraction of the catalyst active sites covered by carbon. Apparently, θ is a function of time and temperature. Thus, the higher the temperature, the more rapid is the drop in the methane decomposition rate. Carbon could be effectively removed from the catalyst surface via gasification reactions with steam, CO_2 and air at temperatures below 850°C. In all cases, the initial catalytic activity toward methane decomposition was practically restored. However, carbon gasification reactions resulted in the production of CO/CO_2 mixtures.

The nature of the methane-metal interaction during decomposition reaction is still much debated. Researchers have found that the activation energy for methane decomposition is lower for the metals with stronger metal-carbon bonds, which correlates with the following order of activity: Fe > Co > Ni.[29] Our experimental data on methane decomposition over alumina-supported Fe, Ni and Co catalysts at 850°C are in a good agreement with the theory. However, at lower temperatures (<700°C), the order of catalytic activity toward methane decomposition changed to Ni > Fe > Co. Apparently, other factors, including hydrogen-metal interaction, play a significant role in methane activation over transition metal catalysts.

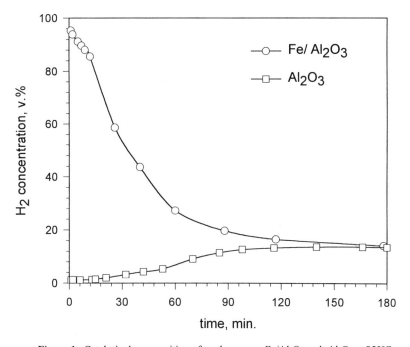

Figure 1. Catalytic decomposition of methane over Fe/Al_2O_3 and Al_2O_3 at 850°C.

No conclusive study is presented in the literature on the mechanism of methane decomposition over metal catalysts. Most likely, a general Langmuir-type mechanism, similar to that suggested for CH_4-D_2 exchange over metal films[30] may be applied to metal-catalyzed methane decomposition reaction:

$$CH_4 + 2* \rightleftarrows CH_3{}^* + H^*$$

$$CH_3 + * \rightleftarrows CH_2{}^* + H^*$$

$$CH_2 + * \rightleftarrows CH^* + H^*$$

$$CH + * \rightleftarrows C + H^*$$

where * is an active site.

The use of carbon-based catalysts offers significant advantages over metal catalysts, as there is no need for the separation of carbon or regeneration of the catalyst: carbon produced builds up on the surface of the original carbon catalyst and could be continuously removed from the reactor (for example, using a fluidized or moving bed reactors). There is a lack of information in the literature on the catalytic properties of various forms of carbon in methane decomposition reaction.

We determined the catalytic activity of various carbon materials (graphite, carbon black, activated carbon, etc.) for the methane decomposition reaction over a wide range of temperatures. Figure 2 depicts the experimental results of the methane decomposition reaction in the presence of different modifications of activated carbons (AC) and microcrystalline graphite in a multi-sectional packed bed reactor at 850°C. The graphite sample demonstrated the lowest activity for the methane decomposition reaction. The rate of methane decomposition over graphite remained practically unchanged during the entire experiment, which indicates that the process promptly reached a steady-state reaction rate controlled by the catalytic activity of carbon produced from methane. The AC catalysts displayed the highest initial activity among the forms of carbon tested. The samples of AC catalysts of different origin and surface area were tested in the methane decomposition reaction at the range of temperatures from 650 to 850°C. The experimental results (see Figure 2) displayed no apparent correlation between AC surface area and their catalytic activity. Two samples of AC with surface areas of 1150 m^2/g (AC-1) and 2000 m^2/g (AC-2) exhibited the highest activity with the initial hydrogen concentrations in the effluent gas reaching up to 95.0 and 93.5 vol%., respectively. This, however, was followed by a gradual drop in the catalytic activity of these catalysts, resulting in a decrease in methane decomposition rate. Thus, over a period of approximately one hour, the AC-1 catalyzed methane decomposition reaction reached a steady-state regime, with the hydrogen concentration close to that of graphite-catalyzed reaction. For AC-2 catalyzed reaction, however, it took much longer (more than two hours) to reach a steady-state regime. The AC catalyst with the highest surface area (2800 m^2/g, AC-3) demonstrated different behavior comparing to the first two samples of AC. The initial maximum hydrogen production rate (with $[H_2]_o=77.5$ vol%) was followed by a rapid (10-15 minutes) drop in hydrogen concentration in the effluent gas, and, after 20 min, a relatively stable methane decomposition process (although with a noticeable decrease in $[H_2]$ over several hours). There were no methane decomposition products other than hydrogen and carbon (traces of ethane and ethylene were detected in the effluent gas after one hour). The amount of carbon produced corresponded to the volume of hydrogen within the experimental margin of error (5%).

The difference in the performance of different forms of carbon can be tentatively explained by the surface structure and size of carbon crystallites. Apparently, carbon materials with surface structures of carbon crystallites close to that of graphite have the lowest catalytic activity toward methane decomposition. This experimental observation is in agreement with the concept discussed in the literature.[31] The total rate of the methane decomposition process is the sum of the rates of carbon nuclei formation and carbon crystallites growth. It was determined that the activation energy of the carbon nuclei formation during methane decomposition (317 kJ/mole) is much higher than the activation energy of the carbon crystallites growth (227 kJ/mole).[31] Thus, in general, the rate of carbon crystallites growth tends to be higher than the rate of carbon nuclei production. The carbon particles produced during methane decomposition over AC catalysts, most likely, have a graphite-like structure. Apparently, in the case of AC-2 catalyst and, particularly, AC-1 catalyst, the rate of carbon crystallite growth exceeds that of nuclei formation. The catalyst surface is rapidly covered with relatively large graphite-like crystallites that occupy active sites and lead to inhibition of the catalytic activity toward methane decomposition. In the case of AC-3 catalyst, for reasons yet to be understood, after the induction period of approximately 15 minutes, the rates of crystallites growth and nuclei formation become comparable, resulting in the quasi-steady-state methane decomposition.

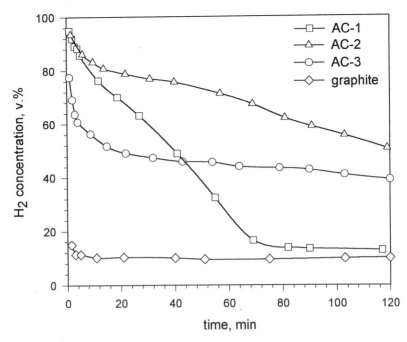

Figure 2. Methane decomposition over different carbon catalysts at 850°C.

The comparison of the methane decomposition reaction in the presence of metal and carbon catalysts reveals some similarities and differences, as shown in Figure 3. The initial hydrogen concentrations in the effluent gas of methane decomposition over Fe and AC-1 catalysts at 850°C are very close and approach the equilibrium value. This indicates that the catalytic activities of fresh AC-1 and Fe catalysts are almost equal at high temperatures. At lower temperatures, however, carbon catalysts are less active than metal catalysts.

It is noteworthy that the steady state concentration of hydrogen in the effluent gases of methane decomposition over Fe-catalyst, AC-1 catalyst and graphite are fairly close. This implies that during a steady-state process, the rate of methane decomposition is determined by the catalytic activity of carbon produced from methane, regardless of the catalyst nature. In order to confirm this assumption, we carried out methane decomposition over alumina at 850°C (see Figure 1). Alumina is practically inert toward methane decomposition, which explains the relatively long induction period of methane decomposition over its surface. As carbon is produced and deposited on the alumina surface, the hydrogen concentration in the effluent gas increases and eventually approaches the steady-state value corresponding to that of Fe- and AC catalysts and graphite. Since carbon produced from methane is the only source of carbon on the alumina surface, this experiment proves that carbon produced from methane controls the rate of methane decomposition during the steady-state regime.

The differences in the temperature dependence and the shape of the kinetic curves for metal- and AC-catalyzed reactions point to the apparent dissimilarities in the mechanism of methane decomposition in the presence of metal and carbon catalysts. The nature of active sites responsible for the efficient decomposition of methane over the fresh surface of carbon catalysts is yet to be understood.

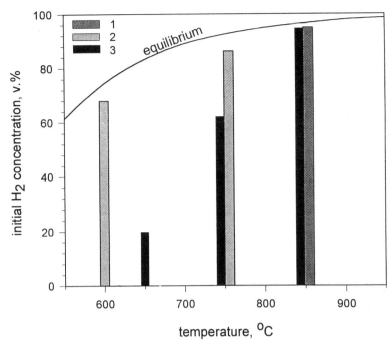

Figure 3. Temperature dependence of methane decomposition reaction in presence of metal and carbon catalysts: 1- Fe/ Al_2O_3, 2- Ni/ Al_2O_3, 3- AC-1.

COMPARATIVE ASSESSMENT OF HYDROGEN PRODUCTION PROCESSES

CO_2 Emissions from Different Hydrogen Production Processes

Several approaches to the comparative assessment of different hydrogen production processes are discussed in the literature.[4,10,32,33] In this work we compared different NG-based hydrogen production processes including SR and PO (with and without CO_2 sequestration), and thermocatalytic decomposition (TCD). The comparison is based on the volumes of H_2 and CO_2 produced per unit of volume of methane consumed. The results are shown in Figure 4.

It is evident that the difference between the amounts of hydrogen produced by SR (without CO_2 sequestration) and TCD (with H_2 as a fuel option) is relatively low (about 37%). On the other hand, SR produces 1 m^3 of CO_2 per each m^3 of methane consumed, whereas TCD is completely CO_2 free. The difference in hydrogen yield from both processes significantly narrows (to approximately19%) when CO_2 sequestration is coupled with the SR process. According to our estimates (see CO_2 Sequestration section), approximately 19% of the thermal energy of methane could be lost during CO_2 sequestration. Somewhat different estimates of the energy loss (15%) was reported in the literature.[33] Thus, due to the energy losses during CO_2 sequestration, the overall SR efficiency significantly decreases and closely approaches that for TCD process. It is evident, also, that a significant amount of CO_2 (up 0.25 m^3 CO_2 per m^3 of CH_4) is produced as a result of CO_2 sequestration associated with the SR process. Thus, we conclude that TCD is the only fossil fuel-based process that shows a real potential to be a completely CO_2-free hydrogen production process.

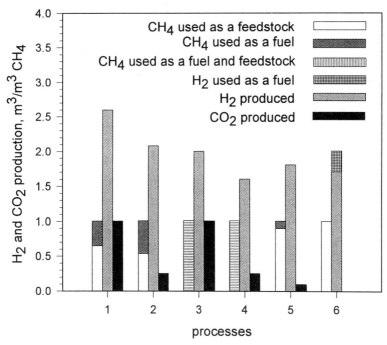

Figure 4. Comparative assessment of hydrogen production processes: 1- SR without CO_2 sequestration, 2-SR with CO_2 sequestration, 3- PO without CO_2 sequestration, 4- PO with CO_2 sequestration, 5- TCD with CH_4 as a fuel option, 6- TCD with H_2 as a fuel option.

Current and Future Markets for Carbon

Currently, the total world production of carbon black is close to 6 million tons per year, with prices varying in the range of hundreds to thousands dollars per ton, depending on the carbon quality.[13] Carbon black has great market potential, both in traditional (rubber industry, plastics, inks, etc.) and new areas such as the metallurgical industry.[34] Carbon black is particularly valuable as a reducing reagent for the production of SiC and other carbides, and as a carbon additive (carburizer) in the steel industry. The carbon black market for these applications in Europe currently approaches 0.5 million tons/year with the prices for the high quality materials reaching $615 per ton. Prices for the good quality carbon black could even reach $1000-4000 per ton.[13] Carbon-based composite and construction materials potentially can absorb a tremendous amount of produced carbon. These materials have remarkable physical properties and can be easily machined, which make them very valuable for construction and lining of technological equipment.

Besides the traditional markets for carbon, some novel applications for the carbon produced via methane decomposition are discussed in the literature. Kvaerner has initiated R&D program to investigate the potential of novel grades of carbon black as a storage medium for hydrogen, and as a feedstock for the production of solar grade silicone.[35] The production of carbon nanotubes and nanofibers via solar thermal decomposition of methane over supported Co and Ni catalysts, respectively, was also reported.[36]

A market for carbon-based materials is continuously growing. However, it is unlikely that all the carbon produced via NG decomposition for mitigating the global warming will be absorbed by the traditional and perspective application areas. In this case, carbon can be stored under the ocean, or in mines and landfill, as discussed in the literature.[33,37] Carbon is perfectly suitable for this purpose: (i) it is chemically inert under ambient conditions, (ii) it has twice the density of water, and (iii) it can be pressed to any shape. No significant energy consumption would be expected with regard to the storage of solid carbon (compared to CO_2 sequestration).

EXPERIMENTAL

Methane (99.99 vol%) (Air Products and Chemicals, Inc.) was used without further purification. Activated alumina, $Ni(NO_3)_2 6H_2O$ and $Fe(NO_3)_3 9H_2O$ (Fisher Scientific) were used without further purification. Alumina-supported Ni and Fe catalysts were synthesized according to the procedures described in the literature.[38] Samples of activated carbon and graphite were obtained from Fisher Scientific and Aldrich, respectively. Preparation of POT photocatalysts and related experimental procedures were described previously.[14]

Methane decomposition experiments were conducted in a 5.0 ml fixed bed quartz microreactor using 0.3 g of catalysts. The catalysts were arranged within the reaction zone in several layers separated with ceramic wool to prevent clogging of the reactor due to produced carbon. The reactor temperature was maintained constant via a type K thermocouple and Love Controls microprocessor. The tubular reactor was made out of alumina and quartz tubings (I.D. 3-6 mm).

Analysis of the products of methane decomposition was performed gas chromatographically: SRI- 8610A (TCD, Ar-carrier gas, silica gel) and Varian-3400 (FID, He-carrier gas, $HysepD_b$), and spectrophotometrically (Spectronic 601).

CONCLUSION

Conventional processes for hydrogen production are among major producers of CO_2 emissions. It has been proposed recently that CO_2 produced in steam reforming or partial oxidation processes could be captured and sequestrated in the ocean or underground. In our work we estimated that the total energy consumption for CO_2 sequestration (CO_2 capture, pressurization, transportation and injection), will most likely exceed 5,000 kJ per kg of sequestrated CO_2. Since about 80% of world energy production is based on fossil fuels, this could potentially result in the production of 0.20-0.25 kg of CO_2 per kg of sequestrated CO_2.

The perspectives of hydrogen production via different methane dissociation processes, including thermal, plasma-assisted, photocatalytic, and thermocatalytic decomposition, are discussed in this paper. The experimental data on thermal, photocatalytic and thermocatalytic decomposition of methane are presented. Thermal homogeneous pyrolysis of methane using a fast pyrolysis reactor resulted in the production of hydrogen, carbon and a wide spectrum of valuable C_2^+ hydrocarbons. Methane was photoactivated by near-UV light in the presence of polyoxocompounds of tungsten, yielding hydrogen and C_2-C_4 olefins. This system could potentially be the basis for the development of solar-driven "green" processes for CO_2-free conversion of natural gas into hydrogen and valuable chemical feedstock. Thermocatalytic decomposition of methane over metal (Fe and Ni) and carbon-based catalysts was investigated. It was demonstrated that at high temperatures (e.g. 850°C), both metal (Fe) and activated carbon catalysts exhibited comparable initial activity toward methane decomposition. The advantage of carbon-based catalysts over metal catalysts is that carbon catalysts potentially do not require regeneration and the process could be arranged in a continuous mode. Some similarities and differences in the mechanism of methane decomposition over carbon and metal catalysts are discussed. It is concluded that thermocatalytic decomposition of NG is the only fossil fuel-based process that shows real potential to be a completely CO_2-free process for hydrogen production.

ACKNOWLEDGMENTS

This work was supported by the U.S. Department of Energy.

REFERENCES

1. B. Cromatry, Modern aspects of steam reforming for hydrogen plants, *Proceedings of the 9th World Hydrogen Energy Conference*, Paris, 13: 2 (1992)
2. N. Nakicenovic, Energy Gases: The Methane Age and Beyond, *IIASA, Working Paper-93-033*, Laxenburg, Austria, 1-13 (1993)
3. K. Blok, R. Williams, R.Katofsky, and C. Hendriks, Hydrogen production from natural gas, sequestration of recovered CO_2 in depleted gas wells and enhanced natural gas recovery, *Energy, 22*: 161 (1997)
4. H. Audus, O. Kaarstad, and M. Kowal, Decarbonization of fossil fuels: hydrogen as an energy carrier, *Proceedings of 11th World Hydrogen Energy Conference*, Stuttgart, Germany, 525 (1996)
5. A. Gritsenko, *Cleaning of Gases from Sulfurous Compounds*, Nedra, Moscow (1985)
6. Energy Information Administration 1998. *Annual Energy Review 1997*, DOE/EIA-0384 (97),U.S. Department of Energy, Washington, D.C. (1997)
7. M. Steinberg, and H. Cheng, Modern and prospective technologies for hydrogen production from fossil fuels, *Proceedings of the 7th World Hydrogen Energy Conference*, Moscow, 699 (1988)
8. M. Steinberg, The Carnol process for CO_2 mitigation from power plants and the transportation sector, BNL 62835, Brookhaven National Laboratory,Upton,NY, (1995)
9. A. Kobayashi, and M. Steinberg, The thermal decomposition of methane in a tubular reactor, BNL-47159, Brookhaven National Laboratory, Upton, NY (1992)
10. E. Shpilrain, V. Shterenberg, and V. Zaichenko, Comparative analysis of different natural gas pyrolysis methods, *Int. J. Hydrogen Energy*, 24: 613 (1999)
11. C. Chen, M. Back, R. Back, Mechanism of the thermal decomposition of methane, *Symp. Industrial and Laboratory Pyrolysis*, (1976)
12. S. Lynum, R. Hildrum, K. Hox, and J. Hugdabl, Kvarner based technologies for environmentally friendly energy and hydrogen production, *Proc. 12th World Hydrogen Energy Conference*, Buenos Aires, 697 (1998)
13. L. Fulcheri, and Y. Schwob, From methane to hydrogen, carbon black and water. *Int. J. Hydrogen Energy, 20: 197 (*1995)
14. N. Muradov, M. Rustamov, and Yu. Bazhutin, Photoactivation of methane and other alkanes using silica-supported polyoxometalates, *Proc. Acad. Sciences USSR*, 312: 139 (1990)
15. J. Pohlenz, N. Scott, Method for hydrogen production by catalytic decomposition of a gaseous hydrocarbon stream, *U. S. Patent No 3,284,161* (UOP) (1966)
16. B. Kim, J. Zupan, L. Hillebrand, and J. Clifford, Continuous catalytic decomposition of methane, *NASA Contractor Report, NASA CR-1662*, NASA, Washington D.C.(1970)
17. M. Callahan, Hydrocarbon fuel conditioner for a 1.5 kW fuel cell power plant. *Proceedings of 26th Power Sources Symposium*, Red Bank, N.J, 181 (1974)
18. M. Pourier, C. Sapundzhiev, Catalytic decomposition of natural gas to hydrogen for fuel cell applications, *Int. J. Hydrogen Energy, 22:429* (1997)
19. K. Ledjeff-Hey, T. Kailk, J. Roes, Catalytic cracking of propane for hydrogen production for fuel cells, *Fuel Cell Seminar*, Palm Springs (1998)
20. T. Koerts, M. Deelen, R. van Santen, Hydrocarbon formation from methane by a low-temperature two-step reaction sequence, *J. Catalysis*, 138: 101 (1992)
21. F. Solymosi, A. Erdohelyi, A. Csereyi, A. Felvegi, Decomposition of CH_4 over supported Pd catalysts *J. Catalysis*, 147: 272 (1994)
22. M. Calahan, Thermo-catalytic hydrogen generation from hydrocarbon fuels. *From Electrocatalysis to Fuel Cells*. G.Sanstede, ed., Battelle Seattle Research Center (1972)
23. N. Muradov, How to produce hydrogen from fossil fuels without CO_2 emission, *Energy and Environmental Progress, Hydrogen Energy and Power Generation*, N. Veziroglu, ed., Nova Science, NY, 93 (1991)
24. N. Muradov, Hydrogen production by catalytic cracking of natural gas, *Proc. 11th World Hydrogen Energy Conference;* Stuttgart, Germany, 697 (1996)
25. A. Pyatenko, M. Nizhegorodova, V. Lipovich, V. Popov, Investigation of the process of catalytic decomposition of methane on d-metals, *Khimiya Tverdogo Topliva*, 23: 682 (1989)

26. J. Rostrup-Nielsen, Equilibria of decomposition reactions of carbon monooxide and methane over nickel catalyst, *J. Catalysis,* 27: 343 (1972)
27. A. Steinfeld, V. Kirilov, G. Kuvshinov, Y. Mogilnikh, A. Reller, Production of filamentous carbon and hydrogen by solar thermal catalytic cracking of methane, *Chem. Eng. Sci.* 52: 3399 (1997)
28. M. Calahan, Catalytic pyrolysis of methane and other hydrocarbons, *Proc. Conf. Power Sources,* 26: 181(1974)
29. E. Shustorovich, The bond-order conservation approach to chemisorption and heterogeneous catalysis: applications and implications, *Advances in Catalysys,* 37:101 (1990)
30. A. Frennet, Chemisorption and exchange with deuterium of methane on metals, *Catal. Rev.- Sci. Eng.,* 10: 37 (1974)
31. A. Tesner, *The Kinetics of Carbon Black Production;* VINITI: Moscow (1987)
32. N. Muradov, CO_2-free production of hydrogen by catalytic pyrolysis of hydrocarbon fuel, *Energy & Fuels,* 12:41 (1998)
33. M. Steinberg, Fossil fuel decarbonization technology for mitigating global warming, *Intern. J. Hydrogen Energy,* 24: 771 (1999)
34. B. Gaudernack, S. Lynum, *Proc. 11th World Hydrogen Energy Conference,* Stuttgart, Germany, 510 (1996)
35. S. Lynum, R. Hildrum, K. Hox and J. Hugdahl, *Proc. 12th World Hydrogen Energy Conference,* Buenos Aires, Argentina, 637 (1998)
36. V. Kirillov, Catalyst application in solar thermochemistry, *Intern. J. Hydrogen Energy,* 66: 143 (1999)
37. N. Muradov, How to produce hydrogen from fossil fuels without CO_2 emission, *Intern. J. Hydrogen Energy,* 18: 211 (1993)
38. J. Richardson, *Principles of Catalyst Development,* Plenum Press, NY (1989)

HYDROGEN PRODUCTION FROM WESTERN COAL INCLUDING CO_2 SEQUESTRATION AND COALBED METHANE RECOVERY: ECONOMICS, CO_2 EMISSIONS, AND ENERGY BALANCE

Pamela Spath and Wade Amos

National Renewable Energy Laboratory
1617 Cole Blvd.
Golden, CO 80401

INTRODUCTION

A collaborative effort to study the possibility of producing hydrogen from low-rank Western coal with an emphasis on CO_2 sequestration and coalbed methane recovery was undertaken by the National Renewable Energy Laboratory (NREL) and the Federal Energy Technology Center (FETC). The purpose of the analysis was to examine the technoeconomic feasibility, CO_2 emissions, and energy balance of several hydrogen production schemes: a reference case, a CO_2 sequestration case, a maximum hydrogen production case, and a hydrogen/power coproduction case. Using their expertise in the field of coal gasification along with their existing models, the researchers at FETC provided NREL with material and energy balance information as well as cost data on the coal gasification and gas clean-up section of the plant. Because of extensive past technoeconomic analysis in the areas of hydrogen production, storage, and utilization, NREL examined the process steps associated with these operations using their previously developed models. NREL then investigated technologies for CO_2 sequestration and coalbed methane recovery and added this information to the analysis. The models were updated and integrated to incorporate the system design details as well as to account for the heat integration of the overall system.

WESTERN COAL ANALYSIS

Wyodak coal from Wyoming was selected as a suitable low-rank Western coal for this study. It is inexpensive to produce and is available in an abundant supply. Additionally, the regulations in Wyoming permit the extraction of coalbed methane, making it attractive for CO_2 sequestration and coalbed methane recovery. Coal would be mined from this region gasified to produce hydrogen then the CO_2-rich off gas would be injected into unmineable coal beds.

Advances in Hydrogen Energy, edited by Padró and Lau
Kluwer Academic/Plenum Publishers, 2000

The elemental analysis and heating value of the selected Wyodak coal[1] is given in Table 1.

Table 1. Wyodak coal analysis for this study

Ultimate Analysis	(Weight %, dry basis)
Carbon	67.6
Oxygen	17.7
Hydrogen	4.8
Nitrogen	1.2
Sulfur	0.8
Ash	7.9
Moisture, as-received (wt%)	26.6
Heat of combustion, higher heating value (HHV), as-received	20,073 J/g (8,630 Btu/lb)

A delivered coal price of $12.85/tonne and a mine mouth price of $5.45/tonne were used in the analysis. Both prices are based on 1997 average coal prices in the state of Wyoming[2].

GASIFIER TECHNOLOGY AND SYSTEM DESCRIPTION

The coal is gasified via the Destec gasifier, which is currently being demonstrated under DOE's Clean Coal Technology Program at the Wabash River Coal Gasification Repowering Project in West Terre Haute, Indiana. This gasifier is a two-stage entrained, upflow gasifier that operates at an exit temperature of 1,038°C (1,900°F) and a pressure of 2,841 kPa (412 psia). The feed is a coal/water slurry containing 53 wt% solids. For hydrogen production, the gasifier must be oxygen blown in order to minimize the amount of nitrogen in the syngas. Nitrogen, like hydrogen, is not strongly adsorbed onto the catalyst in the pressure swing adsorption (PSA) unit, and therefore reduces the hydrogen recovery rate for the same purity.

The synthesis gas leaving the gasifier contains entrained particles of char and ash. Particulate removal is performed using cyclone separators and ceramic candle type hot gas filters. The coal gas is primarily comprised of H_2, CO, CO_2, and H_2O. Since there is less than 0.1 mol% CH_4, reforming of the syngas is not necessary. However, in order to maximize hydrogen production, shift reactors will be needed to convert the carbon monoxide to hydrogen.

Because the syngas from the gasifier contains approximately 1,400 ppm of H_2S, the majority of the sulfur must be removed prior to shift conversion. Hot gas clean up (HGCU) followed by a ZnO bed is the most economical sulfur removal choice because it avoids cooling and reheating the syngas stream, in addition to avoiding condensing out the majority of the steam that is required for shift conversion. The transport desulfurizer technology from the Piñon Pine Project located near Reno, Nevada was used in the HGCU process step. This technology has an absorber/regenerator system where sulfur compounds are absorbed on a zinc oxide based sorbent. When the sorbent is regenerated, SO_2 is captured and converted to sulfuric acid.

Because the gasifier operates at a high temperature, a steam cycle was incorporated into the process design. Stepwise cooling of the synthesis gas produced steam that was used to generate electricity or to fulfill the plant steam requirements. Finally, hydrogen purification is done using a pressure swing adsorption unit.

CO₂ SEQUESTRATION AND COALBED METHANE RECOVERY

Based on data from previous studies[3,4], the analysis assumed that two molecules of CO_2 were injected for every one molecule of CH_4 released from the coalbed. This is based on worldwide data that shows, on average, a little more than twice as much CO_2 can be stored in a methane field, on a volumetric basis, than the amount of CH_4 extracted. The off gas from the hydrogen purification unit containing primarily CO_2 (68 mol%; 93 wt%) must be compressed from 2.6 MPa (372 psi) to a pressure of 3 - 14 MPa (500- 2,000 psi) which is the pressure range generally found in coalbed methane reservoirs[5]. Additionally, the analysis assumed that new wells must be drilled and that they are connected by a CO_2 distribution system. The total length of piping from the hydrogen plant to the wells is presumed to be 100 km, based on a study by Blok et al[6].

HYDROGEN PRODUCTION SCHEMES

In order to compare the economics as well as the overall CO_2 emissions from each schematic studied in this joint venture, a reference case was analyzed. The reference case included only the process steps associated with coal gasification, shift, and hydrogen purification, but none of the steps associated with CO_2 sequestration or coalbed methane recovery. Three other process schemes were examined in this study and compared to the reference case. Figure 1 depicts simplified process flow diagrams for the reference case and the other three schemes (note: the overall heat integration for each scheme is not shown). The top portion of the figure shows the process steps that are the same for each scheme up to hydrogen purification, while the operations inside the dashed boxes represent the steps that differ between the four cases.

Case 1 represents the reference case. The PSA off gas is typically used to fuel the reformer in steam methane reforming plants but, due to the composition of the gasifier syngas, this scenario did not require a reformer. Therefore, the PSA off gas would be emitted directly to the atmosphere or the off gas could be combusted and the flue gas emitted to the atmosphere. Both of these options were examined in the analysis. Case 2 is the reference case with CO_2 sequestration (coalbed methane is not recovered). Case 3 is the maximum hydrogen production scheme. The off gas is injected into the coal seam and a portion of the recovered methane is reformed to produce synthesis gas. This gas is then shifted and purified to produce more hydrogen. Part of the recovered coalbed methane is used to fuel the reformer. An alternative to this would be to use a portion of the PSA off gas as fuel to the reformer prior to injecting the stream into the coal seam. However, because the heating value of the methane stream is much larger than that of the off gas, it was found to be more economical to compress the entire off gas stream and use a portion of the recovered coalbed methane as fuel. Additionally, because the majority of the off gas from the PSA is CO_2, the overall CO_2 emissions with the off gas as a fuel were found to be higher than when methane is used as the fuel. Therefore, the maximum hydrogen production case where a portion of the off gas is used to fuel the reformer is not reported in this paper. Case 4 produces hydrogen from the synthesis gas, with injection of the CO_2-rich off gas into the coalbed, and production of electric power from the recovered methane using a natural gas turbine and steam cycle.

MATERIAL AND ENERGY BALANCE

Table 2 shows the material and energy balance for each case studied. The coal feed rate is the same for each case and the resulting amount of hydrogen does not change except for the

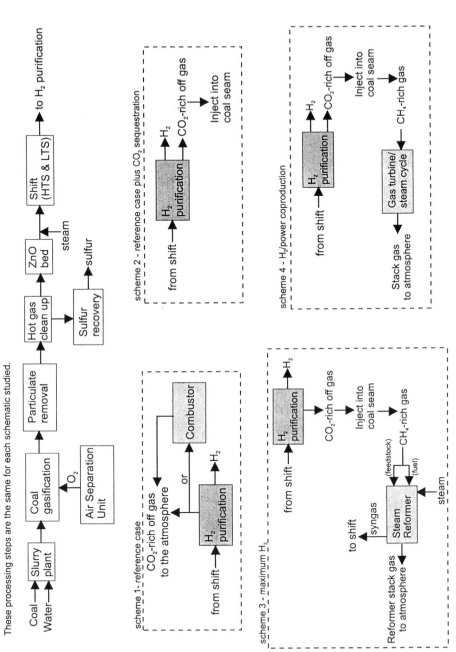

Figure 1. Schematics of the Systems Analyzed.

These processing steps are the same for each schematic studied.

Coal — Slurry plant — Coal gasification — Particulate removal — Hot gas clean up — ZnO bed — Shift (HTS & LTS) — to H₂ purification

Water

O₂ / Air Separation Unit

steam

Sulfur recovery — sulfur

scheme 1 - reference case

from shift — H₂ purification — CO₂-rich off gas to the atmosphere — or — Combustor — H₂

scheme 2 - reference case plus CO₂ sequestration

from shift — H₂ purification — H₂ / CO₂-rich off gas — Inject into coal seam

scheme 3 - maximum H₂

from shift — H₂ purification — H₂ / CO₂-rich off gas — Inject into coal seam — CH₄-rich gas

Steam Reformer — to shift — syngas — (feedstock) — (fuel) — steam

Reformer stack gas to atmosphere

scheme 4 - H₂/power coproduction

from shift — H₂ purification — H₂ / CO₂-rich off gas — Inject into coal seam — CH₄-rich gas — Gas turbine/steam cycle — Stack gas to atmosphere

20

maximum hydrogen production case (Case 3). Additionally, all of the cases examined, except maximum hydrogen production (Case 3), produce some amount of power. The reduction in the amount of power produced by the system between the reference case (Case 1) and the CO_2 sequestration case (Case 2) is a result of the electricity required to compress the CO_2-rich off gas and inject it in the coal seam. The maximum hydrogen production case (Case 3) actually requires more electricity than it produces, mainly due to the compression requirements for the CO_2-rich off gas and the reformer combustion air. There is also an additional electrical load in producing the steam required for reforming the coalbed methane.

Table 2. Material & energy balances at 100% capacity

Case	#	Coal rate (as received) (kg/hr)	Coalbed methane (kg/hr)	H_2 production (kg/hr)	Electricity requirement (MW)	Energy ratio
Reference	1	113,393	0	8,011	-12	0.83[1] 0.58[2]
CO_2 sequestration	2	113,393	0	8,011	-4	0.57
Maximum H_2	3	113,393	47,366	18,739	43	0.65
H_2/power coproduction	4	113,393	36,419	8,011	-241	0.50

Energy ratio defined as (energy out of the system/energy into the system):

$$\frac{(H_2)(HHV_{H2}) + (STM_{ex})(\Delta H_{sh}) + e_{exp} + (off)(HHV_{off})}{(Coal_f)(HHV_{coal}) + (CH_4)(HHV_{CH4}) + e_{imp}}$$

where:
H_2 = hydrogen produced (kg)
HHV_{H2} = higher heating value of hydrogen (GJ/kg)
STM_{ex} = steam produced for export (kg)
ΔH_{sh} = enthalpy difference between incoming water and steam produced for export (GJ)
e_{exp} = exported electricity (GJ equivalents)
off = off gas produced (kg) - reference case only
HHV_{off} = higher heating value of off gas (GJ/kg) - reference case only
$Coal_f$ = coal feed rate (kg)
HHV_{coal} = higher heating value of the coal (GJ/kg)
CH_4 = methane feed rate (kg)
HHV_{CH4} = higher heating value of the methane (GJ/kg)
e_{imp} = imported electricity (GJ equivalents)

[1]This is the energy ratio with an off gas energy credit.
[2]This is the energy ratio without an off gas energy credit.

CO_2 BALANCE

To adequately determine the overall effect of CO_2 for each option studied, the CO_2 balance must incorporate CO_2 emissions in addition to those emitted from the process itself. For example, each case produces electricity, except for the maximum hydrogen production case (Case 3), and for these cases (Cases 1, 2, and 4) a CO_2 emissions credit must be taken for displacing electricity from the grid. Because the maximum hydrogen production case (Case 3) requires some grid electricity, the system must be debited (rather than credited) with CO_2 emissions equivalent to the plant's net electricity requirement. Additionally, for the two options that recover coalbed methane (Case 3 and 4), each of those systems must be credited with CO_2 emissions that are avoided from natural gas production and distribution via today's normal routes of gas and oil wells. Table 3 summarizes the CO_2 emissions for each of the cases examined.

Table 3. CO_2 balance at 100% capacity

Case	#	Overall CO_2 to atmosphere (kg/hr)	Avoided electricity CO_2 (kg/hr)	Avoided natural gas CO_2 (kg/hr)	Electricity CO_2 (kg/hr)	Process CO_2 (kg/hr)
Reference with off gas energy credit	1	195,707	-10,037	N/A	N/A	205,744
Reference without off gas energy credit	1	185,297	-10,037	N/A	N/A	195,334
CO_2 sequestration	2	-3,667	-3,667	N/A	N/A	0
Maximum H_2	3	65,985	N/A	-12,694	35,608	43,070
H_2/power coproduction	4	-109,065	-200,575	-9,760	N/A	101,270

Process CO_2 defined as:
Reference with off gas energy credit = flue gas resulting from combusting the PSA CO_2-rich off gas
Reference without off gas energy credit = CO_2-rich off gas from the PSA directly to the atmosphere
CO_2 sequestration = none
Maximum H_2 = CO_2 in the steam reformer flue gas
H_2/power coproduction = CO_2 in the steam/natural gas combined cycle stack gas

For the reference case (Case 1 - with and without and off gas energy credit), the CO_2 emissions are primarily a result of the hydrogen production process. On a kg/hr basis, the amount of CO_2 emitted in the reference case with an off gas energy credit is only 5.6% higher than the case without the off gas credit. This is due to the fact that the off gas is already rich in CO_2. The overall CO_2 emissions for the CO_2 sequestration case (Case 2) are actually slightly negative instead of zero because of the CO_2 credit for the displaced grid electricity. The hydrogen/power coproduction case (Case 4) also results in a negative amount of CO_2 emissions. This is due to the large credit in CO_2 emissions from displacing a significant quantity of grid electricity. In this analysis, grid electricity was assumed to be the generation mix of the mid-continental United States. According to the National Electric Reliability Council, this includes 64.7% from coal and coal-fired power plants, which generate large quantities of CO_2. Although there are still process emissions from this system, these are overshadowed by the avoided CO_2 emissions. Even though the maximum hydrogen case (Case 3) sequesters most of the CO_2 emissions, some CO_2 is generated when the off gas is burned in the reformer. Additionally, because this case requires a large amount of electricity, the CO_2 emissions from the electricity are nearly equal to the process CO_2 emissions. However, it is not correct to compare the emissions on a per system or a per amount of hydrogen produced basis because many of these cases generate power (refer to Table 2) and all of the cases produced energy in the form of steam. Additionally, for two of the cases (Cases 3 and 4), the additional hydrogen or power is produced from coalbed methane and the energy content of this feedstock must be taken into consideration. To correctly compare the systems, they must be examined on an energy wide basis. Therefore, the CO_2 emissions were divided by the energy ratio of the system and the results can be seen in Figure 2. For comparison, the CO_2 emissions were also plotted assuming that no CO_2 credits or debits were taken for grid electricity and natural gas production and distribution. It is evident that the only case that is greatly affected by this, and would be misrepresented, is the hydrogen/power coproduction case (Case 4).

Hydrogen/power coproduction (Case 4) and CO_2 sequestration (Case 2) are the only cases that result in a negative amount of CO_2 emissions. However, the maximum hydrogen production case (Case 3) does emit significantly less CO_2 than the reference case (Case 1). If the CO_2 emissions were examined per the amount of hydrogen produced from each system, the results would be misleading. Refer to Figure 3, which shows the reference case (Case 1) with the off gas energy credit emitting more CO_2 than the case where the off gas is utilized. Although, in general, the overall trends of Figure 2 and Figure 3 are similar (with the exception of the reference case (Case 1) with the off gas energy credit), the magnitude of the results are different. In Figure 3, the maximum hydrogen production case (Case 3) appears better than actual and the hydrogen/power coproduction case (Case 4) does not look as good as actual.

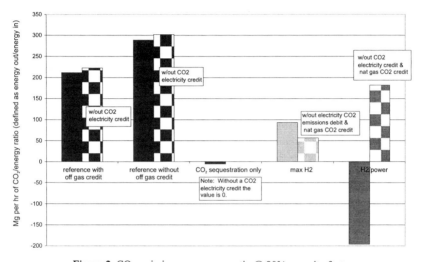

Figure 2. CO_2 emissions per energy ratio @ 90% capacity factor.

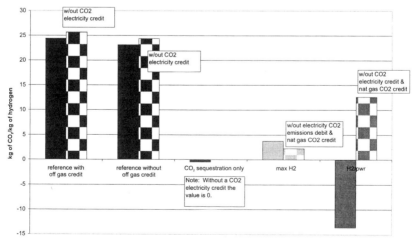

Figure 3. CO_2 emissions per amount of hydrogen produced. For these systems, this is NOT a correct way of looking at emissions.

PLANT GATE COST OF HYDROGEN

To determine the required selling price of hydrogen, a cash flow analysis was performed using an after-tax internal rate of return (IRR) of 15%. Other major assumptions used in the analysis were: equity financing for a 20 year plant life including two years of construction time, a 90% on-stream factor with 50% plant capacity in first year of production, 30% of capital investment is spent in the first year and 70% in the second year, a tax rate of 37%, and ten year straight-line depreciation.

Table 4 summarizes the total installed capital investment for each case. For each system, the majority of the capital cost comes from the oxygen plant (20-24%), and from the coal preparation, gasification, and high temperature cooling sections (20-30%). When the equipment and piping required for CO_2 sequestration (Case 2) is added to the reference case (Case 1), the total installed capital increases by 18.6%. The maximum hydrogen production case (Case 3) and the hydrogen power coproduction case (Case 4) have the largest capital costs because the entire PSA off gas stream is sequestered and either a steam reforming section is added to the plant design to produce more hydrogen from the recovered coalbed methane (Case 3) or a gas turbine/steam cycle section is added to produce power from the recovered coalbed methane (Case 4).

Table 4. Total installed capital cost

Case	#	Total Installed Capital Investment (million, U.S. $)
Reference	1	612
CO_2 sequestration	2	726
Maximum H_2	3	959
H_2/power coproduction	4	929

Figure 4 shows the results for the plant gate hydrogen selling price. The plant gate hydrogen selling price ranged from as low as $9.89/GJ for the maximum hydrogen production case (Case 3) to as high as $18.72 for the CO_2 sequestration case (Case 2). The maximum hydrogen production and hydrogen/power coproduction cases (Cases 3 and 4) were found to be more economical than the reference case (Case 1).

Carbon Tax

By comparing the hydrogen selling price of the reference case (Case 1) with that of the CO_2 sequestration case (Case 2), a carbon tax that would represent a break-even point could be calculated. The hydrogen selling price for the CO_2 sequestration case (Case 2) would be reduced to $17.98/GJ, the reference (Case 1) with off gas credit case cost, if a carbon tax of as little as $13.38/tonne of carbon were mandated. Figure 5 shows how this carbon tax would affect the other cases. The hydrogen selling price for the maximum hydrogen case (Case 3) decreases by $0.17/GJ and the that for the hydrogen/power coproduction case (Case 4) decreases from $13.92/GJ to $12.50/GJ.

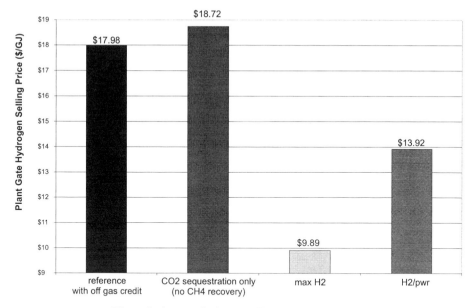

Figure 4. Plant gate hydrogen selling price for all four cases.

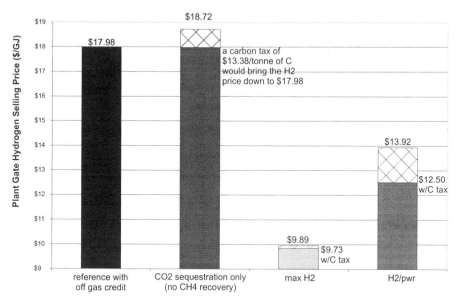

Figure 5. Plant gate hydrogen selling price with a carbon tax.

Sensitivity Analysis

A sensitivity analysis was conducted to examine the effect of varying the base case assumptions for several parameters. The parameters and their changes are shown in Table 5.

Table 5. Variables examined in sensitivity analysis

Variable	Base Value	Sensitivity Analysis Values	
Electricity price	$0.05/kWh	decrease to $0.04/kWh	increase to $0.07/kWh
Coal price	mine mouth price of $5.45/tonne (as received), for all except reference case	delivered price of $12.85/tonne (as received)	
Off gas fuel credit (reference case only)	off gas fuel credit relative to HHV of natural gas	no off gas fuel credit	
Excess steam	credit for selling 100 and 500 psi steam	no credit for excess steam	

Figure 6 shows the change in the plant gate hydrogen selling price for the different parameter values. Note that in a few instances two parameters were changed simultaneously; however, in most instances only one variable was altered at a time. The base case hydrogen selling price for each of the four scenarios is also denoted on the figure. For the reference case (Case 1), the CO_2 sequestration case (Case 2), and the maximum hydrogen production case (Case 3), the parametric studies showed that the hydrogen price is only affected by about +/- $1/GJ. However, as expected, the hydrogen/power coproduction case (Case 4) is greatly affected by the electricity selling price. The only variable that reduces the hydrogen selling price for all four cases is a change in the price of electricity.

DELIVERED PRICE OF HYDROGEN

Thus far, the analysis has examined the plant gate hydrogen selling price. However, often times the user will not be "over the fence". Thus the cost to store and transport the hydrogen must be added to the plant gate price in order to determine the delivered price of the hydrogen. The cost of storing and transporting hydrogen depends on the amount of hydrogen the customer needs and how far their site is from the production facility. Six different storage and transport scenarios, as denoted in Table 6, were examined. These represent six likely scenarios for hydrogen use.

Table 6. Storage and transportation scenarios

Scenario #
1: bulk delivery - 16 km one way
2: bulk delivery - 160 km one way
3: bulk delivery - 1,610 km one way
4: on site consumption - 12 hours of storage; no transport
5: gas station supply - weekly deliveries; driving distance of 160 km round trip; supply multiple stations along the way; hydrogen use of 263 kg/day per gas station
6: pipeline - 3 km to nearest infrastructure; no storage; an additional 160 km pipeline distance for delivery to end user for which the cost is shared by 5 companies

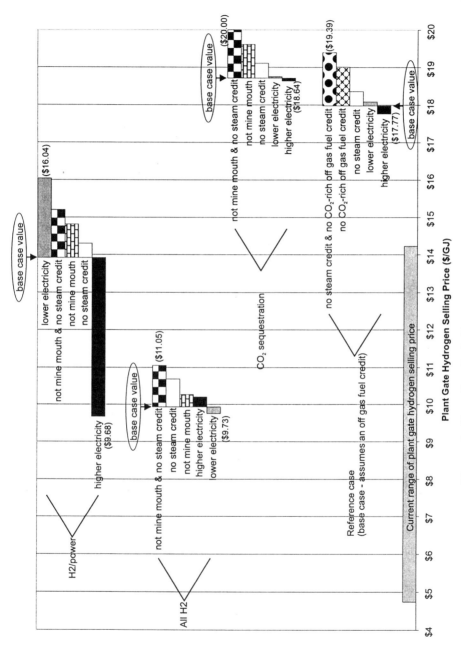

Figure 6. Plant Gate H₂ Selling Price - Sensitivity Analysis.

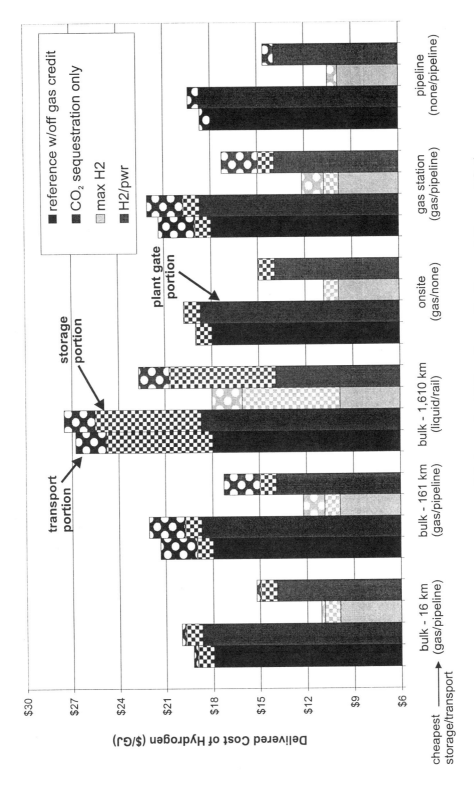

Figure 7. Delivered Cost of Hydrogen for All Four Cases - Six Storage & Transport Options.

In each scenario, the cheapest storage and delivery method was identified based on previous work performed at NREL[7]. The associated incremental costs were added to the hydrogen production price to get the total delivered price of the hydrogen. Figure 7 shows the resulting delivered hydrogen price for the different scenarios. The solid portion of each bar is the plant gate price, the checkered part represents the increase due to storage, and the bubbled section is the contribution from transportation. The cheapest storage and delivery options are denoted in parenthesis. In most instances, compressed gas storage with pipeline transport is the lowest cost choice.

It is apparent that the option of bulk storage with a transport distance of 1,610 km has significantly higher storage and transport costs ($8.8/GJ) than bulk delivery and a shorter transport distance of 16 km, onsite consumption, or pipeline transport of 160 km to the end user shared by 5 companies (around $1/GJ for each of these options). Overall, for the options studied in this analysis, storage and delivery will add $0.6 - $8.8/GJ to the plant gate selling price. The delivered hydrogen selling price is very dependent upon the amount of hydrogen the customer needs and how far their site is from the production facility.

CONCLUSIONS

Four process schemes were evaluated in this study of coal gasification for hydrogen production: 1) a reference case where CO_2 sequestration and coalbed methane recovery are not incorporated, 2) a CO_2 sequestration case, 3) maximum hydrogen production incorporating steam methane reforming of recovered coalbed methane, and 4) hydrogen/power coproduction with the hydrogen being produced from the syngas and the power from the recovered methane. The economics favor sequestering CO_2, recovering coalbed methane, and making hydrogen or power (Cases 3 and 4). However, due to the CO_2 emissions generated from the steam methane reformer, additional hydrogen production via recovered natural gas (Case 3) is not necessarily the most environmentally friendly option from a CO_2 standpoint. Coal fired power plants emit large quantities of CO_2, therefore optimizing hydrogen production with electricity generation, as in Case 4, is a means of lowering the CO_2 emissions from power generation in the U.S. Because of the high temperatures and low CH_4 content in the resulting syngas, coal gasification for hydrogen production does not require a steam methane reforming step, and adding CO_2 sequestration (Case 2), results in almost no CO_2 being emitted to the atmosphere, for minimal additional cost. Mandating a carbon tax would make sequestering the CO_2 economically viable. However, for all of the cases examined in the analysis, it should be noted that there is much debate about the fate of the sequestered CO_2 and its long term environmental effects.

ACKNOWLEDGEMENTS

We would like to acknowledge the Department of Energy's Office of Fossil Energy and Office of Energy Efficiency and Renewable Energy for funding this work. We also acknowledge the assistance of Harold Chambers, Diane Revay Madden, and Denny Smith (FETC) and Walter Shelton (EG&G Technical Services), who provided coal gasification data for this study.

REFERENCES

1. U.S. Department of Energy. (February 1995). *Coal Data: A Reference.* Energy Information Administration, Office of Energy Market and End Use. DOE/EIA-0064 (93). Washington D.C.

2. U.S. Department of Energy. (November 1998). *Coal Industry Annual 1997*. Energy Information Administration, Office of Energy Market and End Use. DOE/EIA-0584(97). Washington D.C.

3. Gunter, W.D.; Gentzis, T.; Rottenfusser, B.A.; Richardson, R.J.H. (September 1996) "Deep Coalbed Methane in Alberta, Canada: A Fuel Resource with the Potential of Zero Greenhouse Gas Emissions." Proceedings of the Third International Conference on Carbon Dioxide Removal. Cambridge, MA..

4. Hendriks, C. (1994). *Carbon Dioxide Removal from Coal-Fired Power Plants*. Kluwer Academic Publishers. Dordrecht, Netherlands.

5. The American Association of Petroleum Geologists. (1994). *Hydrocarbons from Coal*. AAPG Studies in Geology #38. Tulsa, Oklahoma. Edited by Ben Lae and Dudley D. Rice.

6. Blok, K.; Williams, R.H.; Katofsky, R.E.; Hendriks, C.A. (1997) "Hydrogen Production from Natural Gas Sequestration of Recovered CO_2 in Depleted Gas Wells and Enhanced Natural Gas Recovery." *Energy*, Volume 22. Number 2/3. pp. 161-168.

7. Amos, W.A. (1998). *Costs of Storing and Transporting Hydrogen*. National Renewable Energy Laboratory, Golden, CO, TP-570-25106.

UNMIXED REFORMING: A NOVEL AUTOTHERMAL CYCLIC STEAM REFORMING PROCESS

Ravi V. Kumar, Richard K. Lyon, and Jerald A. Cole

General Electric Energy and Environmental Research Corporation
18 Mason
Irvine, California 92618

ABSTRACT

Unmixed Reforming, or UMR, is a novel autothermal cyclic steam reforming process for converting hydrocarbons to hydrogen. The process operates in a three-step cycle that involves heating the reactor, reducing the catalyst to the metallic state, and finally steam reforming. UMR is being developed mainly for small-scale generation of hydrogen where it can economically compete with conventional steam reforming. UMR produces synthesis gas with higher concentrations of hydrogen and lower concentrations of carbon oxides when compared with conventional steam reforming. In addition, the process is more robust than conventional steam reforming, as it can readily be used with feed stocks containing heavy hydrocarbons and sulfur.

INTRODUCTION

Currently, the principal applications for hydrogen and synthesis gas (a mixture of hydrogen and carbon monoxide) are ammonia production, petroleum refining, oxo-synthesis, methanol synthesis, Fischer-Tropsch process and generation of reducing gas for steel production (Rostrup-Nielsen, 1984). The future vision for hydrogen is that it will be widely used as an energy carrier and a fuel and it will be generated from renewable sources (NREL, 1995). Much of the increase in hydrogen use is expected in transportation and electric generation applications (DOE, 1998).

Fuel cells are the primary technology that will advance hydrogen use (DOE, 1998). Fuel cells are important as they are one component of a system that can efficiently produce electricity for many applications (Jacoby, 1999). It is also widely accepted that fuel cells are environmentally friendly (Hirschenhofer, 1997). Low temperature fuel cells, such as polymer-electrolyte-membrane (PEM) fuel cells, are being considered for many applications including electric power generation in commercial and residential buildings, automobile applications and

Advances in Hydrogen Energy, edited by Padró and Lau
Kluwer Academic/Plenum Publishers, 2000

portable power supplies. PEM fuel cells require a continuous supply of hydrogen that is free of contaminants such as carbon monoxide that can damage the fuel cell electrodes.

Hydrogen can be generated from a variety of sources: fossil fuels such as natural gas, naphtha, fuel oils, coal; renewable fuels such as landfill gas; and water by electrolysis. Generating hydrogen from fossil fuels is viewed as a transitional strategy on the pathway to an environmentally responsible, sustainable energy economy (DOE, 1998).

Currently steam reforming is the established and the most economical process for converting natural gas and other light hydrocarbons to hydrogen and synthesis gas (Rostrup-Nielsen, 1993; Twigg, 1996). It is estimated that 76% of all hydrogen produced comes from steam reforming of natural gas (Balthasar, 1984). The most important alternatives are partial oxidation of fuel oil and coal gasification (Rostrup-Nielsen, 1984). Steam reforming is considered economical only at large scales (Che and Bredehoft, 1995). Today, hydrogen is being delivered to small customers by truck or pipeline. In the future, to be cost competitive, hydrogen will need to be produced at the customer site in distributed systems (DOE, 1998).

Recently, there has been significant interest in developing technologies for converting fossil fuels to hydrogen in small-scale systems. This paper presents a novel method of steam reforming hydrocarbon fuels to produce hydrogen.

CONVENTIONAL STEAM REFORMING

In the conventional steam reforming process hydrocarbon feed gas is mixed with steam and passed over a catalyst to produce a mixture of hydrogen and carbon oxides. This product gas from the reformer is commonly referred to as synthesis gas or reformate. Using methane as an example, the process consists of two steps:

$$CH_4 + H_2O + heat = CO + 3 H_2 \qquad (1)$$

$$CO + H_2O = CO_2 + H_2 + heat \qquad (2)$$

Reaction (1) is the primary reforming reaction and is endothermic. Reaction (2) is the water-gas shift reaction and is exothermic. Both these reactions are limited by thermodynamic equilibrium. The overall reaction is endothermic and hence requires that additional fuel be combusted to supply heat. The conventional steam reformer is a fired furnace containing catalyst-filled tubes. The hydrocarbon and steam mixture is processed in the catalyst-filled tubes while external burners heat the tubes. Nickel supported on a ceramic matrix is the most common steam reforming catalyst.

UNMIXED REFORMING: PROCESS DESCRIPTION

The Unmixed Reforming (UMR) process operates by alternately cycling between steam reforming and regeneration. The regeneration step provides the heat required by the endothermic steam reforming reactions utilizing a process known as Unmixed Combustion (UMC).

In the UMC process (Lyon, 1996; Lyon and Cole, 1997) the combustion is carried out by alternately cycling air and fuel over a metal placed in a packed bed reactor. The metal is dispersed on a high surface area support. When air is passed over the packed bed, the metal oxidizes to form metal oxide in a highly exothermic reaction. Since there are no gaseous products for this oxidation reaction the heat of reaction is first directly deposited on the packed bed. A significant portion of the heat remains where it was formed as sensible heat of the bed solids and the remaining moves through the bed as sensible heat of the oxygen depleted air. If regular combustion was used to heat the packed bed, on the other hand, the heat of reaction is first deposited in the gaseous phase and then transferred to the packed bed. The subsequent and separate introduction of a fuel reduces the metal oxide back to metal. The fuel regeneration step is mildly exothermic or mildly endothermic, depending on the choice of fuel and metal. The overall reaction is exothermic and can be used to heat packed beds.

The heat thus generated in-situ over the packed bed is then available for the endothermic steam reforming reactions. Nickel was chosen as the metal for UMR, since it is the most common catalyst for the steam reforming reactions and can be readily oxidized and reduced.

The three steps of the UMR process are shown in Figure 1. The figure portrays the progress of the reaction starting at the left of the figure and moving toward the right. The three steps are referred to as the reforming step, the air regeneration step and the fuel regeneration step. During the reforming step, fuel and steam react over the nickel catalyst to produce hydrogen through conventional steam reforming chemistry. During the air regeneration step, air is passed through the packed bed to oxidize the catalyst. The heat released by the oxidation reaction raises the temperature of the packed bed. In the fuel regeneration step, fuel is introduced to the packed bed, reducing the oxide form of the catalyst back to its elemental form.

The primary reactions that occur in the reactor are as follows:

Reforming Step

$$CH_4 (g) + H_2O (g) = CO (g) + 3H_2 (g) \qquad \Delta H_{rxn} = +205.8 \text{ kJ/mol} \quad (3)$$
$$CO (g) + H_2O (g) = CO_2 (g) + H2(g) \qquad \Delta H_{rxn} = -41.2 \text{ kJ/mol} \quad (4)$$

Air Regeneration Step

$$Ni + 1/2 \, O_2 (g) = NiO \qquad \Delta H_{rxn} = -244.4 \text{ kJ/mol} \quad (5)$$

Fuel Regeneration Step

$$NiO + 1/4 \, CH_4 (g) = 1/4 \, CO_2 (g) + 1/2 \, H_2O (g) + Ni \qquad \Delta H_{rxn} = 43.7 \text{kJ/mol} \quad (6)$$

Reactions (3) and (4) are equilibrium limited, while Reactions (5) and (6) reach completion. An energy balance of the reactions show that oxidation/reduction of one nickel atom produces enough energy to reform approximately one methane molecule to produce four moles of hydrogen.

Figure 2 shows the theoretically predicted gas composition and temperature of the reformate stream produced by the UMR process. The data was calculated by modeling the packed bed as a series of chemical equilibrium reactors. Figure 2 shows that there is a significant drop in temperature during the endothermic reforming step, and as a result, there is significant variation in the product gas composition. At the beginning of the reforming step, the temperature of the packed bed is high. At these high temperatures, the endothermic primary reforming Reaction (3) is favored and the exothermic Reaction (4) is not favored. Hence, the concentration of methane is low and concentration of carbon monoxide is high during the initial stage of the reforming step. The temperature of the packed bed decreases significantly toward the end of the reforming step. The low temperature has the opposite effect on the concentrations of methane and carbon monoxide. The variation in the reformate gas composition can be dampened by adding additional thermal mass in the packed bed.

Addition of Carbon Dioxide Sorbent

The UMR process can be improved by introducing calcium oxide, a carbon dioxide sorbent, into the packed bed. The potential advantages of using calcium oxide as a carbon dioxide sorbent have been previously recognized. The use of calcium oxide to enhance the steam reforming process has been patented by Gluud et al.(1931). More recently Harrison and coworkers (Han and Harrison, 1994) have reported laboratory-scale data for the steam reforming of methane in the presence of calcium oxide.

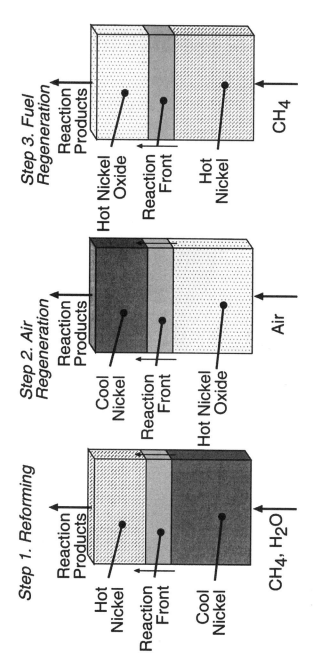

Figure 1. Illustration of the Unmixed Reforming process without calcium oxide. The process consists of three steps: 1) Reforming; 2) Air regeneration; 3) Fuel regeneration. The Unmixed Reforming process cycles through the three steps in the above sequence.

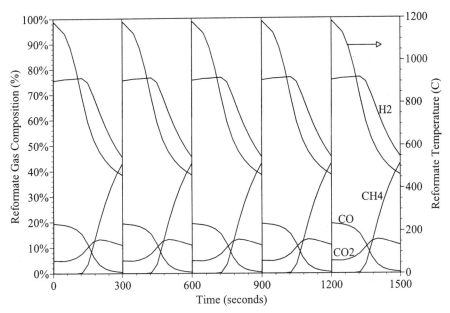

Figure 2. The gas composition and temperature of reformate stream from UMR process without calcium oxide.

In the UMR process, calcium oxide is converted to calcium carbonate during the reforming step since it captures some of the carbon dioxide formed. During the air regeneration step the calcium carbonate is decomposed back to calcium oxide. The UMR process with calcium oxide is shown in Figure 3.

The primary reactions that occur in the reactor are as follows:

Reforming Step

CH_4 (g)+ H_2O (g) = CO (g) + $3H_2$ (g)	ΔH_{rxn}=+205.8 kJ/mol	(7)
CO (g) + H_2O (g) = CO_2 (g) + H_2(g)	ΔH_{rxn}=-41.2 kJ/mol	(8)
CaO + CO_2 (g) = $CaCO_3$	ΔH_{rxn}=-183.0 kJ/mol	(9)

Air Regeneration Step

Ni + 1/2 O_2 (g) = NiO	ΔH_{rxn}=-244 .4 kJ/mol	(10)
$CaCO_3$ = CaO + CO_2 (g)	ΔH_{rxn}=183.0 kJ/mol	(11)

Fuel Regeneration Step

NiO + 1/4 CH_4 (g) = 1/4 CO_2 (g) + 1/2 H_2O (g)+ Ni	ΔH_{rxn}=43.7kJ/mol	(12)

The carbon dioxide sorbent, calcium oxide, serves three functions in the process. First, the adsorption of carbon dioxide by calcium oxide to produce calcium carbonate is an exothermic chemical reaction that also delivers energy to the reforming reactions in the form of chemical potential energy. During the air regeneration step, the process of decomposing the calcium carbonate to calcium oxide largely absorbs heat released by oxidation of nickel, and

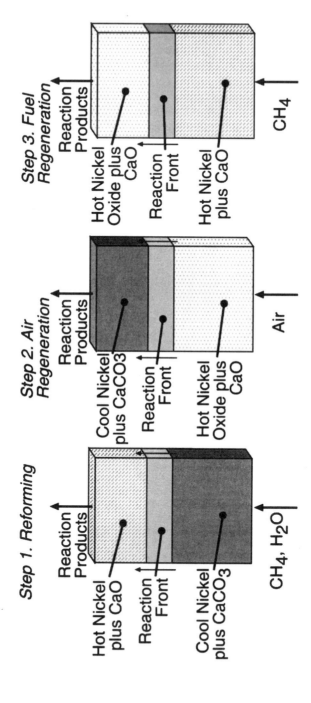

Figure 3. Illustration of the Unmixed Reforming process with calcium oxide. The process consists of three steps: 1) Reforming; 2) Air regeneration; 3) Fuel regeneration. As the nickel is cycled between the oxide and the catalytically active metallic state, the calcium is cycled between the oxide and the carbonate, thus reducing the carbon dioxide concentration in the product hydrogen and effecting the transfer of chemical energy between the regeneration and reforming steps.

this substantially moderates the temperature rise of the packed bed. Much of the enthalpy of nickel oxidation is thereby stored as chemical potential energy in the calcium oxide. This method of using the inter-conversion between calcium oxide and calcium carbonate to supply heat to the reforming process, is a far more efficient means of transferring energy to the reforming process. Furthermore, reducing the temperature rise during regeneration reduces parasitic heat losses from the reactor and promotes process efficiency. The heats of reaction that are presented along with Reactions (7), (8) and (9) indicate that the reforming step is now nearly thermoneutral.

Second the sorbent provides additional "thermal mass" to transfer sensible heat from the regeneration step to the reforming step of the process.

Finally, the presence of the solid-phase sorbent material improves the equilibrium of Reactions (7) and (8). By removing carbon dioxide from the products of the steam reforming process, equilibrium is shifted toward greater hydrogen production, reduced carbon monoxide and carbon dioxide concentrations, and increased fuel conversion.

Figure 4 shows the theoretically predicted gas composition of the reformate stream for the UMR process with calcium oxide. The addition of calcium oxide substantially moderates the temperature drop during the reforming step and hence reduces the changes in reformate gas composition.

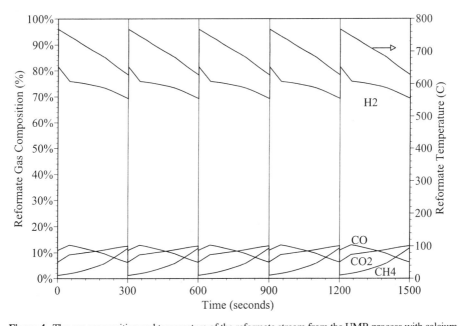

Figure 4. The gas composition and temperature of the reformate stream from the UMR process with calcium oxide.

A COMPARISON BETWEEN CONVENTIONAL STEAM REFORMING AND UNMIXED REFORMING

There are several differences between conventional steam reforming and UMR: cost of reformer, heat transfer limitations, reaction equilibrium limitations, effectiveness of catalyst and feed stock limitations.

First, consider the cost. The conventional steam reforming process typically requires that the process gas be heated over the catalyst to temperatures of 870°C. This requires that the metal walls of the tubes be heated to 900°C (Tindall and King, 1994). In order to withstand the high temperatures, the tubes are made of high-alloy nickel chromium steel. The tubes are expensive and account for a large portion of the reformer costs (Adris et al., 1996). Increasing the furnace temperature to enhance the heat transfer is not feasible, since even a slight increase in tube-wall temperature may result in a serious decline in the expected tube-life time (Adris et al, 1996). On the other hand, the UMR reactor does not require the high temperature tubes as it is an internally heated packed bed. However, UMR is a cyclic process and hence requires additional process instrumentation.

Second, the efficiency of conventional reformer decreases at small scales due to heat transfer limitations and parasitic heat losses. The transfer of heat from the combustion products to the reactants is an inherently inefficient process, and in any practical system, especially for smaller scales, it is not possible to transfer all of the energy released by combustion into the process being heated. In UMR the heat transfer is more efficient as the packed bed is internally heated. The improvements in heat transfer to the packed bed are expected to result in a higher process efficiency.

Another limitation is dictated by equilibrium constraints for the reforming Reactions (1) and (2), i.e., they do not reach completion. As a result, the reformate stream contains unconverted methane, carbon monoxide and carbon dioxide. Typically, at lower temperatures (600°C) the amount of carbon monoxide in the reformate gas is reduced to 2%, however the amount of methane increases to 28%. At high temperatures (900°C), the amount of methane in the reformate is reduced to 2%, however the amount of carbon monoxide increases to 20%. In UMR reaction equilibrium is shifted favorably by the carbon dioxide sorbent.

In conventional reforming the reactor's high aspect ratio (L/D) requires that large catalyst particles be used in order to avoid excessive pressure drop. Due to intra-particle diffusional limitations in large particles, 95% of the catalyst is not utilized (Adris et al., 1996). In UMR, the use of low aspect ratio reactors allows the use of small catalyst particles and as a result the effectiveness factors of the catalyst are higher.

The range of hydrocarbons that can be used in conventional steam reformers is limited by three factors: the feed's tendency to form coke, the feed's sulfur content, and the locally available supply of the feed.

In conventional steam reforming the catalyst is always kept under strongly reducing conditions. Consequently any coke, which forms on the catalyst, must be removed by steam gasification, a slow process, or it will accumulate and deactivate the catalyst. Another major problem of formation of coke is the severe consequences of mal-distribution of heat, i.e., the danger of tube metal failure due to overheating caused by the formation of hot spots (Adris et al., 1996). Even though commercial steam reforming catalysts typically include components that catalyze steam gasification, conventional steam reforming technology is still restricted to naphtha and other hydrocarbons that are lighter than naphtha (Tindall and Crews, 1995). Feeds with a tendency to form coke, e.g., logistics fuel, can not be used (the U.S. Department of Defense has decided to simplify the logistics of military operations by having all equipment run on a single fuel. This logistics fuel is to have specifications similar to diesel and jet fuel). Czernik et al. (1999) have studied a process in which biomass is pyrolyzed into an ash-free mixture of gases and liquid organic matter and the latter is steam reformed into hydrogen. The initial high conversion of the feed to hydrogen decreased with time due to coke accumulation.

The formation of coke is not a problem for UMR since any coke that is formed is burnt off during the air regeneration step. This allows the use of UMR with diesel/logistics fuel and possibly with biomass pyrolysis liquids, though the latter has not yet been demonstrated.

The presence of sulfur in the feed is also a problem for conventional steam reforming. If the catalyst is not sulfur tolerant, the sulfur will deactivate it (Rostrup-Nielsen, 1984). If the catalyst is "sulfur tolerant", i.e., able to tolerate small amounts of feed sulfur, the sulfur leaves the process as hydrogen sulfide and is likely to cause unacceptable problems downstream. Furthermore many major natural gas resources are "sour," i.e., contain a large percentage of hydrogen sulfide.

UMR is tolerant of sulfur in the feed, which makes it a more robust process when compared with conventional steam reforming. In UMR, some of the sulfur reacts with both nickel and calcium oxide during the reforming step, but is rejected as sulfur dioxide during the air regeneration step. With respect to using UMR to process sour natural gas, it is interesting to note that, at 850°C, a carbon-to-water ratio of twice stoichiometric, and 8 bar, the heat of reaction for steam reforming of methane to an equilibrium mixture of carbon monoxide, carbon dioxide and hydrogen is +200 kJ/mol of converted methane. During UMR any hydrogen sulfide in the feed is converted to sulfur dioxide and water vapor with a heat release of -519 kJ/mol of sulfur. Thus if the sour gas has a molar ratio of hydrogen sulfide to methane, of 0.38 or greater, it is, in theory, possible to steam reform all the methane to hydrogen without using any for process heat.

Conventional steam reforming is also restricted to applications in which the size of the locally available supply of hydrocarbon fuel is adequate. Much of the world's total natural gas resources are what is called "static gas," i.e., natural gas resources which are individually so small and so remotely located that they can not be economically pipelined to market. In theory this static gas could be reformed into synthesis gas which could then be made into readily shipped liquids. As discussed by Che and Bredehoft (1995), the minimum size for an economically viable steam reformer based on conventional technology is 5,000,000 standard cubic feet (scf) of hydrogen per day. To give such a minimum size steam reformer a 20-year useful life, the local natural gas resource would need to be relatively large. Studies of the economics of UMR indicate that the process will be satisfactory in small-scale applications.

EXPERIMENTAL

The UMR process was studied in a pilot scale experimental system. A simplified schematic of the pilot scale system is shown in Figure 5. The pilot scale system consists of two packed bed reactors. The system was designed to produce 100 standard liters per minute of hydrogen, which is sufficient to generate 10 kW of electricity using a PEM fuel cell.

The reactors are packed with a mixture of the catalyst and the carbon dioxide sorbent. The catalyst is Nickel supported on calcium aluminate and the carbon dioxide sorbent is dolomite. The particle sizes for the catalyst and dolomite are around 1 to 7 mm.

Switching valves are used to deliver the feed streams for the different steps of the UMR process. The system can continuously produce hydrogen by operating the two reactors in four steps:

- Reactor A is in the steam reforming step while reactor B is in the air regeneration step
- Reactor A is in the steam reforming step while reactor B is in the fuel regeneration step
- Reactor A is in the air regeneration step while reactor B is in the steam reforming step
- Reactor A is in the fuel regeneration step while reactor B is in the steam reforming step

A back pressure regulator is used to set the pressure in the reactor which is in the reforming step to 1-7 bar. The pressure during the regeneration step is around 1-2 bar. On-line gas analyzers were used to measure the concentrations of carbon monoxide, carbon dioxide and oxygen. The concentration of hydrocarbons was measured using an on-line flame ionization analyzer. The readings from the on-line analyzers were verified using gas chromatography with a thermal conductivity detector.

RESULTS & DISCUSSION

Reaction equilibrium analysis

The reforming step of the UMR process was analyzed by performing reaction equilibrium calculations. The familiar McBride/Gordon code (McBride & Gordon, 1996) was used to perform the calculations.

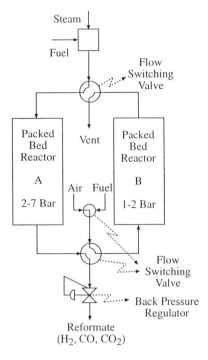

Figure 5. Schematic of experimental system.

The reaction equilibrium for the reforming step of the UMR process depends on four variables: temperature, pressure, steam-to-carbon molar ratio in the feed stream and calcium-to-carbon molar ratio. It was assumed that the steam to carbon molar ratio is three for the following analysis.

The parameters used to assess the UMR performance are: methane conversion, and hydrogen-to-carbon-monoxide molar ratio in the reformate stream.

Figure 6 shows methane conversion during the reforming step as a function of temperature for four different pressures and four different calcium-to-carbon molar ratios. As shown in Figure 6, higher conversion of methane is favored at higher temperatures and lower pressures. This is because higher temperatures and lower pressures favor Reaction (7), an endothermic reaction through which there is an increase in the total number of moles. Figure 6 also shows that as the calcium-to-carbon molar ratio is increased, the methane conversion increases. This is because at higher calcium-to-carbon molar ratios, Reaction (9) is favored, shifting the equilibrium of Reactions (7) and (8) to higher methane conversions. Increasing the calcium-to-carbon molar ratio above unity does not significantly affect the UMR reaction equilibrium. However, due to intra-particle diffusion limitations, higher calcium-to-carbon ratios are used in the UMR process.

Figure 7 shows the opposite effect of temperature and pressure on the ratio of hydrogen-to-carbon-monoxide in the reformate stream, i.e., lower temperatures and higher pressures favor higher hydrogen-to-carbon-monoxide ratios. This is because lower temperatures favor exothermic Reactions (8) and (9), and higher pressures favor Reaction (9), shifting the equilibrium of Reaction (8) toward higher conversion of carbon monoxide to carbon dioxide. Figure 7 also shows that as the calcium-to-carbon molar ratio is increased the carbon monoxide concentration in the hydrogen product stream decreases. This is because at higher calcium-to-carbon molar ratios Reaction (9) is favored, shifting the equilibrium of Reaction (8) to higher conversion of carbon monoxide to carbon dioxide.

The figures discussed above indicate that the optimum temperature for the reforming step in the UMR process is 700-850°C, the optimum pressure is less than 7 bar and the optimum calcium-to-carbon molar ratio during the reforming step is around one. Under these optimum conditions the methane conversion varies from 82 to 100% and the hydrogen-to-carbon-monoxide molar ratio in the reformate stream varies from 6 to 100.

Experimental validation

Pilot plant experiments, using diesel fuel, have demonstrated product gas hydrogen concentrations typically averaging 70+ percent and as high as 85 percent under certain conditions, with the balance primarily methane, carbon monoxide and carbon dioxide. Figure 8 shows the experimental data collected over a one hour period of the pilot plant operation during which the average purity of hydrogen in the reformate stream was maintained at 70%. The reforming step was conducted for 300 seconds. At the beginning of the reforming step the hydrogen concentration is low since the nickel oxide in the packed bed is still being reduced to nickel by fuel. As the reforming step progresses the temperature of the packed bed decreases and as a result the methane concentration in the reformate stream increases and the carbon monoxide concentration decreases.

The pilot scale unit was also used to study the UMR process using natural gas as the feed-stock. The experimentally measured gas composition of the reformate stream during one cycle of the UMR process is shown in Figure 9. The gas composition data calculated using the model showed similar trends (see Figure 4).

PEM fuel cells require that the concentration of carbon monoxide be reduced to less than 10 ppm, and hence a high hydrogen to carbon monoxide ratio is preferable in the reformate stream. A comparison of hydrogen to carbon monoxide molar ratios for different hydrogen producing technologies is shown in Figure 10. The UMR process was able to generate a hydrogen product gas with an average hydrogen to carbon monoxide molar ratio in excess of 10. Partial oxidation and autothermal reforming produce a reformate stream with hydrogen to carbon monoxide molar ratios ranging from less than 2 to about 3, while conventional steam reforming attains a ratio of about 5.

Bench scale experiments using diesel fuel have shown that for typical fuel sulfur concentrations there is no detectable sulfur in the reformate to at least the sub-ppm level. In the bench scale experiments, when sulfur was added to diesel fuel at a concentration of 2000 ppm by weight only 5 ppm hydrogen sulfide was detected in the reformate. In pilot scale experiments using diesel fuel, the sulfur concentration in the reformate stream was around 15 ppm. If the sulfur was not captured by the pilot-scale reactor, the sulfur concentration in the reformate stream would have been 37 ppm.

Due to the manner in which fuel is oxidized in UMR, the byproduct exhaust gases contain no oxides of nitrogen. The nitrogen oxide concentrations were below the detection limit of the gas analyzer, roughly 0.03 ppm.

Process stability

The exothermic and the endothermic reactions in the UMR process can be balanced by selecting the appropriate cycle times, fuel, steam and air flow rates. The repeatability in the gas composition data from cycle to cycle, as shown in Figure 8, is an indication of the stability of the UMR process.

There is some mixing of air and fuel during switching between steps. However, this occurs in the presence of an ignition source (hot catalyst) which avoids the possibility of forming a potentially explosive mixture in cooler regions of the system. In the pilot-scale program additional precautions were taken to ensure that the dimensions of the system were too small to permit detonation of even hydrogen-air mixtures.

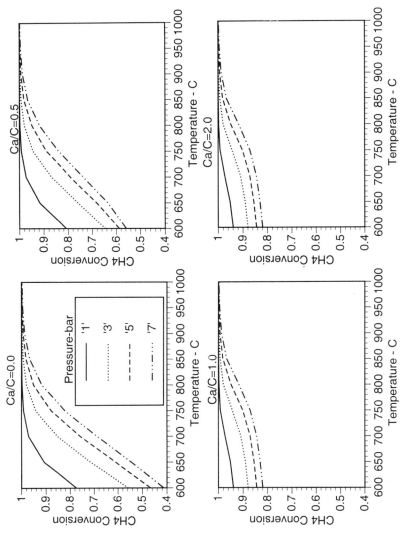

Figure 6. Effect of temperature, pressure and calcium-to-carbon molar ratio on methane conversion. The calcium-to-carbon molar ratio is for the entire reforming step (steam-to-carbon molar ratio of 3).

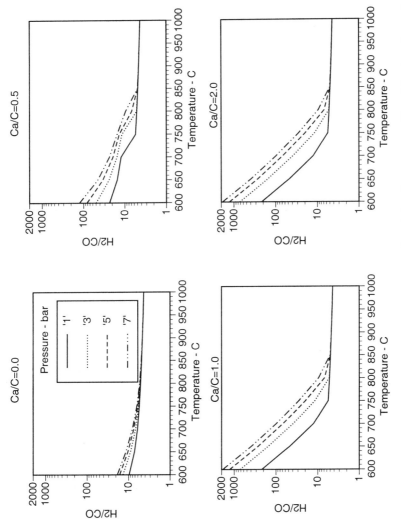

Figure 7. Effect of temperature, pressure and calcium-to-carbon molar ratio on H2-to-CO molar ratio. The calcium-to-carbon molar ratio is for the entire reforming step (steam-to-carbon molar ratio of 3).

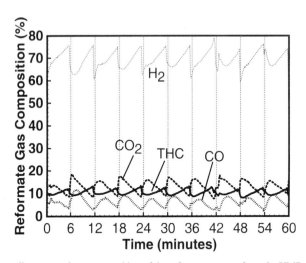

Figure 8. Experimentally measured gas composition of the reformate stream from the UMR process with diesel fuel as feed.

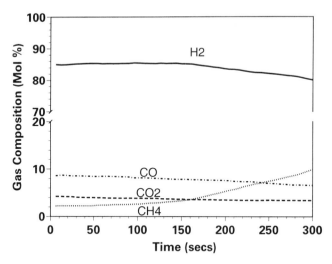

Figure 9. Experimentally measured gas composition of the reformate stream from the UMR process with natural gas as the feed.

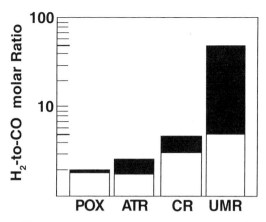

Figure 10. A comparison of hydrogen to carbon monoxide molar ratios for different processes for producing hydrogen from fossil fuels (POX = partial oxidation; ATR = autothermal reforming; CR = conventional steam reforming; UMR = unmixed reforming).

ACKNOWLEDGMENTS

The authors would like to acknowledge the support of U.S. Defense Advanced Research Project Agency (contract #DAAH01-95-C-R162) and U.S. Department of Energy (contract #DE-FC02-97EE50488).

REFERENCES

Adris, A.M., Pruden, B.B., Lim, C. J. and Grace, J.R., 1996, On the Reported Attempts to Radically Improve the Performance of the Steam Methane Reforming Reactor, Canadian J. Chem. Eng., 74, p177.

Balthasar W., 1984, Hydrogen Production and Technology: Today, Tomorrow and Beyond, Int. J. Hydrogen Energy, 2, 649.

Che, S.C. and Bredehoft, R. L., Preprints, 1995, Div. of Petroleum Chemistry, American Chemical Society, Sep 20-25.

Czernik, S., French, R., Feik, C., and Chornet, E., 1999, Preprints, Div. of Fuel Chemistry, American Chemical Society, 44, #4, 855-858.

DOE Technology validation plan., 1998, Hydrogen Program, Document # DOE/GO-10099-740.

Gilles, E.A., 1980, Chem. Eng. Prog., 76, 88.

Gludd,W., Keller, K., Schonfelder, R. and Klempt, W., 1931, Production of hydrogen, U.S. Patent 1,816,523.

Han, C. and Harrison, D. P., 1994, Simultaneous shift reaction and carbon dioxide separation for the direct production of hydrogen, Chem. Engg. Sci., 49, p5875.

Hirschenhofer, J.H., 1997, Fuel Cell Status:1996, IEEE AES Systems Magazine, 3, 23.

Jacoby, M., 1999, Chem. Eng. News., 77, p31.

Lyon, R.K., 1996, Method and Apparatus for Unmixed Combustion as an Alternative to Fire. U.S. Patent No. 5,509,362.

Lyon, R.K. and Cole, J.A., 1997, Unmixed Combustion for Efficient Heat and Mass Transfer in Chemical Process Systems, American Flame Research Committee Fall Symposium Proceedings.

McBride and Gordon, 1996, Computer Program for Calculation of Complex Chemical Equilibrium Compositions and Applications, NASA Reference Publication 1311.

National Renewable Energy Laboratory, 1995, Advanced hydrogen production technologies, DOE/GO-10095-065.

Rostrup-Nielsen, J.R., 1984, "Catalytic steam reforming" in "Catalytic science and technology," J.R. Anderson and M. Boudard, Ed., Vol. 4, Springer, Berlin, p1.

Rostrup-Nielsen, J.R., 1993, Production of synthesis gas, Catalysis Today, 18, p305-324.

Tindall, B.M. and Crews, M.A., 1995, Alternative technologies to steam-methane reforming, Hydrocarbon Processing, 11, p75.

Tindall, B.M. and King, D. L., 1994, Designing steam reformers for hydrogen production, Hydrocarbon Processing, 7, p69.

Twigg, M.V., 1996, Catalyst Handbook, 2nd edition, Wolfe, England, p. 225.

FUEL FLEXIBLE REFORMING OF HYDROCARBONS FOR AUTOMOTIVE APPLICATIONS

J. P. Kopasz, R. Wilkenhoener, S. Ahmed, J. D. Carter, and M. Krumpelt

Chemical Technology Division
Argonne National Laboratory
9700 South Cass Ave
Argonne, IL

INTRODUCTION

Fuel cell vehicles offer many advantages when compared to internal combustion or battery-powered electric vehicles. Advantages over the internal combustion engine (ICE) include the potential for higher fuel efficiency and lower emissions. The advantages over a battery-powered vehicle include an improved driving range and shorter refueling times. The fuel efficiency of a fuel cell vehicle is expected to be about twice that for current internal combustion engines and the overall energy consumption (fuel chain and vehicle) is expected to be lower than that of battery-powered vehicles.[1] Emission levels are expected to meet the Super Ultra Low Emission Vehicle Standard, much lower than those from current ICEs.

The ideal fuel for the low-temperature proton exchange membrane (PEM) fuel cells being considered for automotive applications is hydrogen. Currently, the infrastructure for hydrogen refueling is lacking, and hydrogen storage technologies available for onboard storage provide a decreased driving range compared to gasoline and ICE technology. However, it is apparent that the commercial success of a fuel cell vehicle will be tied to the availability of a refueling infrastructure.[2] In other words, it will be difficult to sell hydrogen-powered fuel cell vehicles without first investing in a hydrogen refueling infrastructure. However, it will be difficult to convince investors to build a hydrogen infrastructure if there are no commercial vehicles to use it.

A solution to this "chicken or the egg" dilemma is to provide an onboard reformer to convert a hydrocarbon fuel into a hydrogen-rich gas for utilization by the fuel cell. This strategy could help introduce fuel cell cars to the marketplace earlier and smooth the transition from internal combustion engine to fuel cell-powered vehicles. Hydrocarbon fuels can use the existing infrastructure for refueling and provide a higher hydrogen density than current hydrogen-storage technologies.

Currently, hydrogen is produced industrially from natural gas using a steam reforming process. A similar process can be used for onboard conversion of natural gas or higher hydrocarbons to hydrogen-rich product gases. However, onboard reforming presents several unique challenges, which include size and weight limitations and the need for rapid startup and the need to be responsive to demand. In addition, since the fuel to be used for onboard reforming is still to be determined, the reformer should be fuel-flexible. These demands suggest a partial oxidation or autothermal reforming process.[3,4,5]

The overall partial oxidation reaction of alkanes can be written as follows:

$$C_nH_{2n+2} + xO_2 + (2n-2x)H_2O = nCO_2 + (3n+1-2x)H_2 \qquad (1)$$

This can be thought of as an exothermic oxidation reaction combined with an endothermic steam-reforming reaction.

Oxidation $\qquad C_nH_{2n+2} + 1/2(3n+1)O_2 = nCO_2 + (n+1)H_2O \qquad (2)$

Steam Reforming $\qquad C_nH_{2n+2} + 2nH_2O(g) = nCO_2 + (3n+1)H_2 \qquad (3)$

The overall partial oxidation catalytic reforming can be exothermic or endothermic. The main factor determining the heat balance for the reaction is the oxygen-to-carbon ratio. The thermal neutral point (where enthalpy of the reaction is zero) varies from an x/n ratio of 0.23 for methanol to 0.37 for iso-octane. It is advantageous to run in the exothermic region, and at a low x/n ratio to maximize the yield of H_2. The experiments reported here were performed with an x/n ratio of 0.5, except for methanol and ethanol experiments, which were performed at a lower x/n (0.32 and 0.25, respectively) to compensate for the oxygen already present in the alcohol.

There is some debate about which hydrocarbon fuel is optimal for fuel cell systems. Methanol and ethanol are available commodity chemicals and have numerous advantages as fuel (e.g., water soluble, renewable), and methanol is easy to reform. Gasoline and diesel have advantages over the alcohols, including existing refueling infrastructures and higher energy density. However, they are blends of different kinds of hydrocarbons and are more difficult to reform. Synthetic Fischer-Tropsch fuels have also been proposed as fuels for fuel cell systems.[6] A reformer that can deal with all these fuels will have advantages. Argonne National Laboratory has developed a partial oxidation catalyst that can convert a broad range of alcohols and hydrocarbon fuels into a hydrogen-rich product gas.[7,8] Tests in microreactors have shown high conversion and hydrogen yields from methane, methanol, ethanol, 2-pentene, cyclohexane, iso-octane, hexadecane, gasoline, and diesel.[9]

EXPERIMENTAL APPROACH

The catalytic partial oxidation reforming of hydrocarbon fuels ($C_nH_mO_p$) was demonstrated in a series of experimental activities, beginning with microreactor studies and progressing to bench-scale reactor investigations. Since gasoline and diesel fuels are blends of different types of hydrocarbons, the studies were initially carried out with representatives of the different types of hydrocarbons present in these fuels, such as iso-octane for paraffins and toluene for aromatics. A schematic of the micro-reactor apparatus is shown in Figure 1. The liquid fuel and water are pumped through separate vaporizer coils. Nitrogen is added as a purge and for use as an internal standard. The vapors then combine with oxygen to form the reactant mix that flows into the microreactor. For higher boiling fuels such as diesel, the fuel bypasses the vaporizer and is injected directly at the front of the catalyst bed. The reactor is a 10.6-mm-ID tube packed with ~2 g of catalyst. The temperature is controlled by a surrounding furnace. Samples are drawn from the product stream and analyzed with a gas chromatograph or gas chromatograph–mass spectrometer. Temperatures and pressures are measured above and below the catalyst bed. In order to facilitate comparisons between the various fuels, microreactor tests were generally run with a fixed carbon-to-oxygen ratio and a slight excess of water with respect to Reaction (1) [$H_2O > (2n - 2x - p)$]. In bench scale reactor tests air is used as the oxidant.

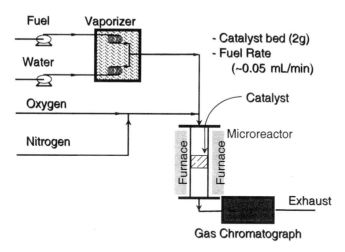

Figure 1. Schematic diagram of microreactor apparatus

RESULTS AND DISCUSSION

Microscale Tests

Hydrogen production from the partial oxidation reforming of several of the fuels tested is shown in Table 1. The table lists the hydrocarbon feed, the reactor temperature at which complete conversion of the hydrocarbon was achieved, and the percentages of hydrogen, carbon monoxide, and carbon dioxide measured in the product gas (after correction for the nitrogen and water present). The last three columns list the calculated percentages of the gases that would exist at equilibrium at the different temperatures. Using the Argonne catalyst, these hydrocarbons were completely converted at temperatures of less than 700°C. Comparison between the experimental and equilibrium gas compositions indicates that the use of the catalyst allows for lower carbon monoxide and slightly higher hydrogen concentrations than might be achieved at equilibrium at these temperatures.

Table 1. Experimental Product Gas Composition Compared with Equilibrium Compositions Calculated for Given Feed Mixture and Experimental Temperature

Fuel	Temperature for Complete Conversion °C	Experimental (%, dry N_2-free)			Equilibrium (%, dry N_2-free)		
		H_2	CO	CO_2	H_2	CO	CO_2
Iso-octane	630	60	16	20	57	20	19
Toluene	655	50	8	39	49	23	26
2-Pentene	670	60	18	22	56	21	21
Ethanol	580	62	15	18	62	18	16
Methanol	450	60	18	20	60	19	17

Table 1 also indicates that the two alcohols (methanol and ethanol) are reformed at substantially lower temperatures than the other species. Of the fuels considered for partial oxidation reforming, alcohols should be considerably easier to reform since they are already partially oxidized.

There is a substantial difference in the temperature at which complete conversion of the hydrocarbons was achieved, even between the two alcohols studied. Figure 2 illustrates the temperature dependence of hydrogen production from the partial oxidation reforming of methanol and ethanol using the Argonne partial oxidation catalyst. The two alcohols behave quite differently. Methanol was easily reformed throughout the temperature range investigated.

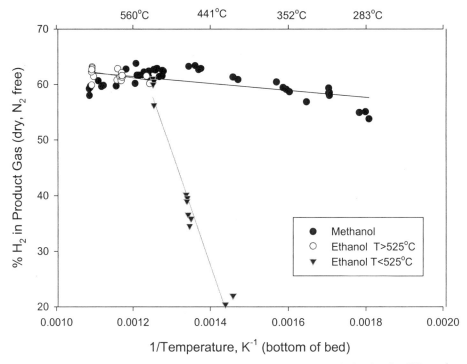

Figure 2. Hydrogen production from partial oxidation reforming of alcohol fuels using the ANL catalyst.

Methanol yielded a product gas that had hydrogen concentrations in the 55 to 65% range (on a dry N_2-free basis), with the product gas varying only slightly in composition with changes in temperature. The temperature dependence of ethanol reforming was quite different, with the product composition varying strongly with temperature in the range of 410 to 525°C. At 470°C, the product gas from ethanol reforming contained only about 35% hydrogen, compared to about 60% hydrogen for methanol reforming at the same temperature. For temperatures greater than 525°C, hydrogen production from ethanol was almost identical to that from methanol. However, the maximum (theoretical) hydrogen yield for the reactant mix with methanol was 70.2%, while that for ethanol was 71%. Defining hydrogen selectivity as the measured hydrogen yield divided by the maximum hydrogen yield, we find slightly lower hydrogen selectivity for ethanol in the high-temperature region (88%) than for methanol (91%) because of the formation of methane during ethanol reforming. The ethanol reformate contained substantial amounts of methane over the entire temperature range investigated; the reformate contained about 5% methane at 640°C, while the methanol reformate contained an order of magnitude less methane at this temperature.

The effect of temperature on hydrogen production from reforming methane, iso-octane, and 2-pentene is shown in Figure 3. These hydrocarbons were more difficult to reform than the alcohols, as expected. Temperatures of nearly 650°C were needed to approach 60% hydrogen in the product gas. Hydrogen selectivities of 90% for methane and 88% for iso-octane and 2-pentene were obtained. The hydrogen concentrations at a given temperature were similar, regardless of the hydrocarbon. The data for methane, iso-octane, and 2-pentene all fall on the same regression line, as shown in Figure 3. Hydrocarbon length over the range of C1 to C8 and the presence of unsaturation appear to have little effect on the product composition.

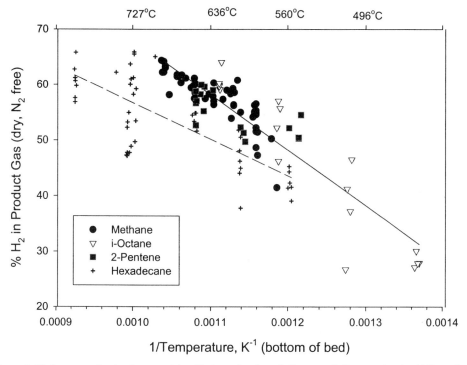

Figure 3. Hydrogen production from partial oxidation reforming of alkanes and alkenes using the ANL catalyst.

Increasing the hydrocarbon length to C16 (hexadecane) was observed to lead to lower hydrogen content in the product gas. Hexadecane reforming showed a temperature dependence similar to that observed for methane, iso-octane, and 2-pentene (i.e., similar slope in Figure 3). However, higher temperatures were needed to reach comparable hydrogen content in the product gas. This appears to be due to increased methane formation from hexadecane. At 650°C, the product gas from hexadecane reforming contained 6% methane (dry, N_2-free), while that from iso-octane reforming contained only about 2% methane.

The iso-octane and hexadecane experiments also highlight some potential difficulties in reforming longer hydrocarbons. The product gas from iso-octane and hexadecane reforming contained small amounts of benzene. This is illustrated in Table 2, which compares the output from thermal reforming (no catalyst present) and catalytic reforming (catalyst present) of iso-octane. Benzene was present in the product gas at concentrations of about 0.1-0.2% for temperatures from 600 to 725°C when the catalyst was present; however, no benzene was detected in the product gas when the catalyst was absent. This suggests that benzene is formed as a side product during the reforming of C8 hydrocarbons.

Table 2. Effect of Catalyst on Product Composition

	Product Composition at 725°C, % (dry, N_2 free)					
	H_2	O_2	CH_4	CO	CO_2	C_6H_6
No Catalyst	0.0	10.2	17.7	39.6	26.1	0.0
Catalyst	57.9	0.0	4.3	19.2	18.1	0.2

The reforming of cyclohexane (a possible benzene precursor) and toluene (a benzene derivative) were also investigated. The hydrogen production from partial oxidation reforming of cyclohexane and toluene did not exhibit the same temperature dependence as did that from reforming the alkanes and the alkene discussed previously. The product composition was much less sensitive to temperature for these cyclic compounds (see Figure 4). Hydrogen selectivity was still high, but slightly lower for toluene (82%) than for cyclohexane (91%) and the acyclic alkanes (88%). The lower temperature-dependence suggests that a different mechanism may be controlling the reforming of cyclohexane and toluene.

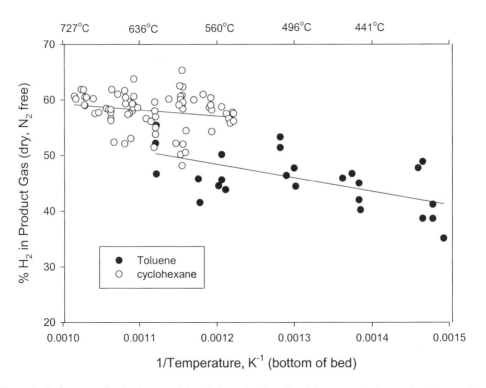

Figure 4. Hydrogen production from partial oxidation reforming of cyclohexane and toluene using ANL catalyst.

The results from reforming toluene suggest that if benzene, xylene, or other aromatics are formed at low levels as side products in the reforming of branched hydrocarbons, they can be reformed. Increasing the temperature will increase the conversion of these aromatics to hydrogen, but not as effectively as it increases the rate of conversion of alkanes to hydrogen.

The ability of the Argonne catalyst to reform several types of hydrocarbons suggested that this catalyst would be able to reform commercial petroleum fuels. However, commercial fuels offer additional challenges to reforming. Gasoline and diesel fuels both contain sulfur, which may poison the catalyst. In addition, gasoline contains a high percentage of aromatics, which may promote coking, while diesel fuel contains long-chain hydrocarbons, which may prove difficult to crack. Both fuels contain additives such as detergents, lubricants, and anticorrosives, which may adversely affect catalyst performance.

Autothermal reforming of commercial gasolines has been performed with the Argonne catalyst. The product composition from reforming of a commercial premium gasoline obtained from a local filling station is shown in Figure 5. Hydrogen concentrations in the product gas approaching 60% on a dry N_2-free basis were obtained at temperatures $\geq 700°C$. The slope of the plot of inverse temperature versus hydrogen content in the product gas is similar to that found for the reforming of methane, iso-octane, and 2-pentene, suggesting a similar mechanism. However, methane concentrations in the product gas are higher than what we have seen from reforming the simple hydrocarbons. Commercial gasoline contains sulfur, a known catalyst poison, at levels up to 1000 ppm. However, no change was observed in the product composition after 40 h of intermittent gasoline reforming, indicating the Argonne catalyst may have some resistance to sulfur poisoning.

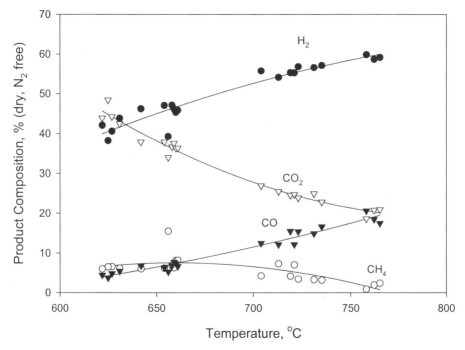

Figure 5. Product distribution from partial oxidation reforming of premium gasoline using the ANL catalyst.

The catalytic partial oxidation reforming of diesel fuel with the Argonne catalyst was also investigated. Results are shown in Figure 6. The diesel fuel was injected directly into the catalyst bed because it contains very high-boiling materials. The direct injection results in minor pressure surges when droplets of fuel hit the catalyst bed, and is responsible for the scatter observed in the data. Diesel fuel proved slightly more difficult to reform than gasoline. Methane was observed in higher concentrations in the product gas from diesel reforming than in the product from gasoline reforming. Higher reforming temperatures were needed to decrease the methane and increase the hydrogen in the product gas. At 850°C, a product gas with hydrogen concentrations >50% on a dry N_2-free basis was obtained (see Figure 6). Again, although the diesel fuel contains sulfur impurities, no degradation in catalyst performance was observed over the duration of the test (8 h).

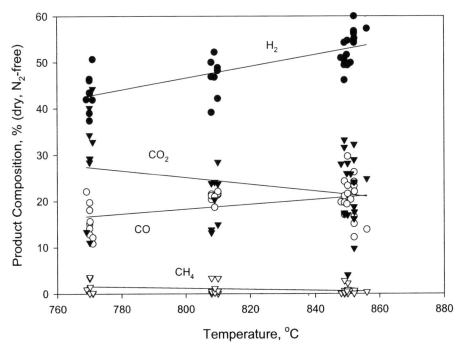

Figure 6. Product distribution from partial oxidation reforming of diesel fuel using the ANL catalyst.

Although the Argonne partial oxidation catalyst appears to exhibit some resistance to sulfur poisoning, sulfur impurities in the fuel will affect the fuel cell itself, and additional fuel processing must be performed to remove sulfur from the feed stream to the fuel cell. This has led some researchers to suggest the use of sulfur-free synthetic fuels as automotive fuel cell fuels. We have performed some initial tests reforming synthetic sulfur-free Fischer-Tropsch fuels using the Argonne partial oxidation catalyst. The fuel was highly saturated, containing > 95% alkanes. The average composition of the fuel was $C_{7.01}H_{15.9}$. The composition of the product gas is illustrated in Figure 7. Over 55% hydrogen was obtained at 755°C; however, substantial methane (approximately 5%) was present. Further optimization of the reaction conditions should lead to decreased methane and higher hydrogen content.

Bench-Scale Tests

Reforming tests were performed on an integrated bench-scale reactor. The reactor consists of a sulfur removal catalyst, the partial oxidation catalyst, and a water-gas shift catalyst. The overall volume is 7 L. The reactor is autothermal, that is, it is self-heated after ignition. Figure 8 illustrates the results for a step change in feed rates (iso-octane for fuel, air as the oxidant). The reactor was running at a steady hydrogen output of over 30 L/min. The fuel feed rate was then increased, resulting in an increase in the hydrogen output. The increase in output was quite rapid, reaching 80% of maximum within 1 minute after the increase in fuel flow. Results are expected to improve further with optimization. This integrated reformer has been demonstrated with natural gas, methanol, ethanol, and gasoline. Hydrogen outputs corresponding to more than 5 kW(e) have been obtained.

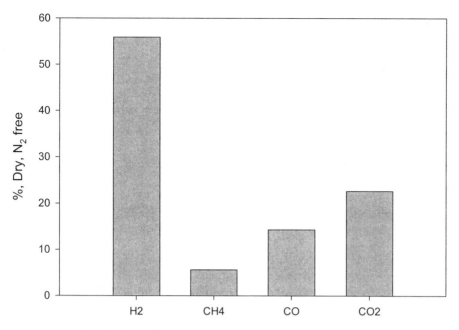

Figure 7. Product Distribution from Reforming Synthetic Sulfur-Free Fuel Using the ANL Catalyst.

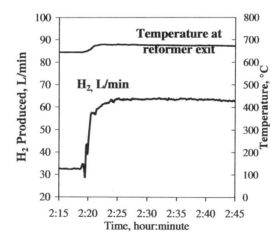

Figure 8. Results from bench-scale partial oxidation reformer demonstrating the rapid response to a demand for increased power.

CONCLUSIONS

Microreactor tests indicate that the Argonne partial oxidation catalyst is fuel-flexible and can reform conventional (gasoline and diesel) and alternative (ethanol, methanol, natural gas) fuels to hydrogen-rich product gases with high hydrogen selectivity. Alcohols are reformed at lower temperatures (<600°C), while alkanes and unsaturated hydrocarbons require higher temperatures (630-670°C). Cyclic hydrocarbons and aromatics have also been reformed at relatively low temperatures; however, a different mechanism appears to be responsible for their reforming. Complex fuels like gasoline and diesel, which are mixtures of a broad range of hydrocarbons, require temperatures >700°C for maximum hydrogen production.

ACKNOWLEDGMENTS

This work was supported by the U.S. Department of Energy, Office of Advanced Automotive Technologies, in the Office of Transportation Technologies, under contract No. W-31-109-ENG-38. The authors would like to thank Syntroleum Corporation for providing the synthetic hydrocarbon fuel.

REFERENCES

1. K. H. Hauer, O Duebel, H. Friedrich, W. Steiger, and J. Quissek, Powertrain International, Vol. 2, p. 23 (1999).
2. R. L. Espino, "Fuels Processing for PEM Fuel Cells - Fuel Industry Perspective," Annual Automotive Technology Development Contractors' Coordination Meeting- PNGV Workshop on Fuel Processing for PEM Fuel Cells, October 23-27, 1995, Dearborn, MI.
3. R. Kumar, S.Ahmed, M. Krumpelt, and K. M. Myles, "Reformers for the Production of Hydrogen from Methanol and Alternative Fuels for Fuel Cell Powered Vehicles," Argonne National Laboratory Report ANL-92/31 (1992).
4. W. L. Mitchell, J. M. Bentley, and H. J. Thijssen, " Development of Fuel Processors for Transportation and Stationary Fuel Cell Systems," Program and Abstracts, Fuel Cell Seminar, Orlando, FL (1996).
5. J. C. Frost and J. G. Reinkingh, "Methanol POX Reforming," Preprints of the Annual Automotive Technolpogy Development Contractors' Coordination Meeting- PNGV Workshop on Fuel Processing for PEM Fuel Cells, October 23-25, 1995, Dearborn, MI (1995).
6. S. Ahmed, J. P. Kopasz, B. J. Russell, and H. L. Tomlinson, Proceedings of the 3rd International Fuel Cell Conference, Nov. 30 – Dec. 3, 1999, Nagoya, Japan
7. R. Kumar, S. Ahmed, and M. Krumpelt, "Rapid-Start Reformer for Methanol and Fuel Cell Vehicles," Electric and Hybrid Vehicle Technology 96, pp. 123-127 (1996).
8. S. Ahmed, R. Doshi, R. Kumar, and M. Krumpelt, "Gasoline to Hydrogen: A New Route for Fuel Cells," Electric and Hybrid Vehicle Technology '97, pp. 77-88, (1997).
9. J. P. Kopasz, R. Wilkenhoener, S. Ahmed, J. D. Carter, and M. Krumpelt, Division of Fuel Chemistry, Preprints of Symposia, 218th American Chemical Society National Meeting, New Orleans, LA, Aug 22-26,1999, p. 899.

THE PRODUCTION OF HYDROGEN FROM METHANE
USING TUBULAR PLASMA REACTORS

Christopher L. Gordon, Lance L. Lobban, and Richard G. Mallinson

Institute for Gas Utilization Technologies
School of Chemical Engineering and Materials Science
University of Oklahoma
Norman, Oklahoma 73019

INTRODUCTION

There are many factors that are changing and shaping the fuel and energy industries of the future. Environmental, political, economic, and availability issues are just some of these factors. With environmental regulations becoming stricter, the emission of greenhouse gases is a major concern. With the decrease in oil resources, there is a need for other sources of fuel and chemical production.

Large quantities of hydrogen are used as a feedstock in the manufacturing of ammonia, methanol, and a variety of other petroleum processes. The synthesis of methanol occurs via the following reactions:

$$CO + 2H_2 \leftrightarrow CH_3OH, \qquad \Delta H_{298} = -91 \text{ kJ mol}^{-1} \qquad [1]$$

$$CO_2 + 3H_2 \leftrightarrow CH_3OH + H_2O, \qquad \Delta H_{298} = -50 \text{ kJ mol}^{-1} \qquad [2]$$

$$CO + H_2O \leftrightarrow CO_2 + H_2, \qquad \Delta H_{298} = -41 \text{ kJ mol}^{-1} \qquad [3]$$

Synthesis gas (CO and H_2) production contributes a large fraction, approximately 60%, of the cost of methanol. The synthesis gas for methanol used to be manufactured by coke gasification, but now is almost exclusively produced by steam reforming of natural gas.

The production of hydrogen from methane has received a lot of research interest over the last decade. There are many good reasons for the conversion of methane, the principle component of natural gas, to other products. Natural gas is a very abundant resource with reserves throughout the world. Methane, with its 4:1 hydrogen to carbon ratio is also an excellent source for hydrogen. Hydrogen is projected to play an important role as a source of energy in the years to come. There will be a large increase in hydrogen demand as it becomes a general-purpose energy source for space heating, electrical power

Advances in Hydrogen Energy, edited by Padró and Lau
Kluwer Academic/Plenum Publishers, 2000

generation, and as a transportation fuel (Balasubramanian, et al. 1999). Also, hydrogen is a clean burning fuel that can be stored as a liquid or a gas, and distributed by means of a pipeline (Armor 1999).

Catalytic steam reforming of methane is currently the primary means of hydrogen production. About 50% of all hydrogen produced worldwide is produced from methane, with 40% of that coming from the steam reforming of methane. It can be seen from Table 1 that the steam reforming of methane has the lowest CO_2 impact compared to other fossil fuels.

Table 1. Variance of CO_2 coproduction with different hydrocarbon feedstocks (Scholz 1993)

H_2/CO_2	Technology
4.0	Steam methane reforming
3.2	Steam pentane reforming
3.0	Partial oxidation of methane
1.7	Partial oxidation of heavy oil
1.0	Partial oxidation of coal

The production of hydrogen from steam methane reforming results from the following two reversible steps:

$$CH_4 + H_2O \leftrightarrow CO + 3H_2, \qquad \Delta H_{298} = 206 \text{ kJ mol}^{-1} \qquad [4]$$

$$CO + H_2O \leftrightarrow CO_2 + H_2, \qquad \Delta H_{298} = -41 \text{ kJ mol}^{-1} \qquad [5]$$

with the overall reaction written as follows:

$$CH_4 + 2H_2O \leftrightarrow CO_2 + 4H_2, \qquad \Delta H_{298} = 165 \text{ kJ mol}^{-1} \qquad [6]$$

While the reforming step [4] does not produce any carbon dioxide, it is the water gas shift reaction [5] that produces the carbon dioxide while removing the carbon monoxide and, in return, yielding another hydrogen molecule. Thermodynamically, the methane steam reforming process is favorable at high operating temperatures and low pressures due to the endothermic reaction and the increase in moles. The high operating temperature requires an intensive energy input to maintain these high temperatures. In addition, it is necessary to operate the system with excess steam in order to reduce the formation of carbon deposits. This in itself is an extra cost with the increase in equipment size.

Therefore, it is desirable to produce synthesis gas more economically. Cold plasmas or "non-equilibrium" plasmas have been shown to activate methane at temperatures as low as room temperature (Liu, et al. 1998). A cold plasma is characterized by high electron temperatures, while the bulk gas temperature can remain as low as room temperature, decreasing or eliminating the heat transfer energy requirements. It is the highly energetic electrons that allow for the conversion of methane that otherwise would not be feasible at these temperatures. In this paper, we discuss the use of our electrical discharge system to convert methane into hydrogen, acetylene, and carbon monoxide. A valuable feature of this system is the very low concentrations of carbon dioxide and water formed.

EXPERIMENTAL

In general, the experimental apparatus is similar to the system that has been described previously (Liu, et al. 1998). The feed gases consisted of a combination of methane, oxygen, hydrogen, and helium. Helium was only used in initial experiments and for characterization studies of the catalyst. The feed gas flowrates were controlled by Porter mass flow controllers, model 201. The feed gases flowed axially down the reactor tube. The reactor was a quartz tube with a 9.0 mm O.D. and an I.D. of either 4.5 mm or 7.0 mm. The configuration of the reactor can be seen in Figure 1.

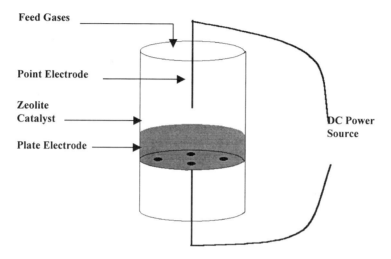

Figure 1. Point-plane dc reactor configuration with catalyst bed.

The reactor's electrode configuration consists of a point to plane geometry, meaning that the top electrode is a wire point electrode and the bottom electrode is a flat plate that also serves as a support for the catalyst. The top electrode is positioned concentrically within the reactor, and the gap between its tip and the plate is 8.0 mm. The catalyst is loaded from the top onto the flat plate electrode. A stainless steel wire cloth is placed between the electrode and catalyst in order to prevent the catalyst from falling through the holes on the electrode plate. Another stainless steel wire cloth is placed on top of the catalyst in order to prevent the movement of the catalyst. Due to the electrostatic nature of the zeolites and the plasma itself, this top screen is necessary to keep the catalyst bed uniform. The preparation and characterization of these zeolites has been discussed elsewhere (Liu, et al. 1996; Marafee, et al. 1997). The dc corona discharge is created using a high voltage power supply (Model 210-50R, Bertan Associates Inc.).

As mentioned before, this system operates at low temperatures. A furnace around the reactor is used to heat the system to the desired temperature. The temperature is measured by an Omega K-type thermocouple that is attached to the outside of the reactor near the catalyst bed. The temperature measured on the outside has been calibrated against the internal temperature of the reactor, and has been discussed elsewhere (Liu, et al. 1996; Marafee, et al. 1997). However, when the operating temperature is below 373 K it is necessary to use cooling air to control the temperature since the plasma itself does heat the gas to some extent. The system pressure varies depending on the experiment. The pressure is controlled using a GO back pressure regulator.

The product gases are passed through a dry ice/acetone bath that allows for any condensable organic liquids to be separated from the product gases. It should be noted that the dc system does not produce any liquids, including water. The effluent gases can be analyzed on-line by either a gas chromatograph or a mass spectrometer. The gas chromatograph is a CARLE series 400 AGC (EG&G) gas chromatograph equipped with a hydrogen transfer system to quantify the hydrogen; a HayeSep column to quantify carbon dioxide, ethane, ethylene, and acetylene; and a molecular sieve column to quantify oxygen, nitrogen, methane, and carbon monoxide. Also, a MKS mass spectrometer is used for on-line analysis of the products and for temperature programmed oxidation of carbon deposited on the catalyst.

RESULTS AND DISCUSSION

The plasma discharge, or cold plasma, is very effective in the conversion of methane at low temperatures. Figure 2 shows the effect of residence time on the conversion of methane. Reactors with two different cross sectional areas were employed to study bypassing of the feed gases around the plasma discharge because the streamer discharges only occupy a fraction of the reaction volume with the streamers moving around the plasma zone. The conditions in these two different reactors were identical except for the inside diameter of the reactors. The depth of the catalyst bed remained constant in both reactors; 0.1 grams and 0.04 grams were used in the large and small reactors, respectively. The original reactor had a 7.0 mm I.D., while the smaller reactor had a 4.5 mm I.D.

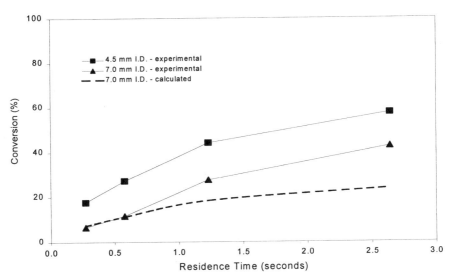

Figure 2. Methane conversion versus residence time in two different cross sectional area reactors. Feed composition, 4/1 CH_4/O_2; gas temperature, 298 K; power, 5.2 watts. Calculated line is for 7.0 mm reactor based on 4.5 mm reactor data.

The fractional conversion is higher in the smaller cross sectional reactor for all residence times. As expected, the conversion increases as the residence time increases for both reactors. The data in Figure 2 show that the larger reactor has a lower conversion than the smaller reactor at the same residence time. This is indicative of bypassing, as the volume average rate of reaction is lower. This suggests that, in contrast to a typical homogenous reactor, there is a zone of higher reaction rate in proximity to the visible discharges and a zone of lower reaction rate, perhaps around the periphery of the reactor. However, if the added peripheral volume of the larger tube (from the difference between the cross sectional areas of the large and small tubes) were assumed to have no reaction, the dashed line drawn on Figure 2 would be expected. This is in agreement with the data at the shorter residence times, but the fact that the experimental data is higher at longer residence times suggests that some level of radial mixing occurs that reduces bypassing at lower velocities. As tube size is made larger such mixing would be desirable to increase the volumetric averaged reaction rate.

Hydrogen, acetylene, and carbon monoxide are the primary products of the conversion of methane in the dc plasma system. In addition, small amounts of ethane, ethylene, and carbon dioxide are produced. Of the C_2 products, acetylene accounts for 90%, while ethylene and ethane comprise 6% and 4%, respectively. Carbon dioxide is less than 0.2% of the effluent gas. No measurable amount of water is produced in the system.

The selectivity and yield of hydrogen, acetylene, and carbon monoxide can be seen in Figures 3-5. The definitions of conversion, selectivity, and yield for this system are as follows:

$$CH_4 \text{ conversion} = (\text{moles of } CH_4 \text{ consumed/moles of } CH_4 \text{ introduced}) \times 100\%$$

$$O_2 \text{ conversion} = (\text{moles of } O_2 \text{ consumed/moles of } O_2 \text{ introduced}) \times 100\%$$

$$\text{Selectivity of } C_2H_6 = 2 \times (\text{moles of } C_2H_6 \text{ formed/moles of } CH_4 \text{ consumed}) \times 100\%$$

$$\text{Selectivity of } C_2H_4 = 2 \times (\text{moles of } C_2H_4 \text{ formed/moles of } CH_4 \text{ consumed}) \times 100\%$$

$$\text{Selectivity of } C_2H_2 = 2 \times (\text{moles of } C_2H_2 \text{ formed/moles of } CH_4 \text{ consumed}) \times 100\%$$

$$\text{Yield of } C_2 \text{ hydrocarbons} = CH_4 \text{ conversion} \times \Sigma(\text{selectivities of } C_2H_2, C_2H_4, C_2H_6)$$

$$\text{Selectivity of } H_2 = 0.5 \times (\text{moles of } H_2 \text{ formed/moles of } CH_4 \text{ consumed}) \times 100\%$$

$$\text{Yield of } H_2 = CH_4 \text{ conversion} \times \text{selectivities of } H_2$$

$$\text{Selectivity of } CO = (\text{moles of } CO \text{ formed/moles of } CH_4 \text{ consumed}) \times 100\%$$

$$\text{Yield of } CO = CH_4 \text{ conversion} \times \text{selectivities of } CO$$

In general, the yields of the products follow the same trend as that of the methane conversion. The selectivities of the respective products did not vary much with either reactor size or residence time. Figure 3 shows that the selectivity of hydrogen has little variance between the different reactor sizes and residence times. The largest variance is 10%, with experimental uncertainty accounting for at least 2-3% of this.

The acetylene selectivity, seen in Figure 4, shows a trend of increased selectivity toward acetylene with residence time for both reactor geometries. In addition, the 4.5 mm reactor has a higher selectivity towards acetylene than the 7.0 mm reactor, and the difference in selectivity gets larger as residence time increases. Therefore, it seems that the higher the conversion, whether achieved by altering the reactor size or residence time, the higher the selectivity towards acetylene. Carbon monoxide selectivity, shown in Figure 5, goes through a maximum at the intermediate residence times.

The above results were for a feed composition of 80% methane and 20% oxygen. In order to reduce carbon monoxide production impact, that might later have to be emitted as carbon dioxide, a study was done in which the feed composition was 66% hydrogen, 32% methane, and 2% oxygen (2/1 H_2/CH_4 with 2% oxygen). Previous studies showed that, with little or no oxygen in the feed, it is necessary to have a H_2/CH_4 ratio above one in order to achieve a stable plasma discharge in a non-oxidative environment. If the H_2/CH_4 ratio is less than one, the discharge will change from the streamer discharges to an unfavorable arc discharge. It is presumed that the 20% oxygen in the first set of experiments and the hydrogen in the second set of experiments play an important role in cleaning the catalyst, which in return plays an important role in the stability of the streamer discharges. It has also been determined from previous studies that 2-2.5% oxygen is still needed for the non-oxidative conditions. Concentrations lower than 2% have instability problems, while concentrations above 2.5% yield stable discharges, but do not increase methane conversion. In addition, the concentrations of oxygen above 2.5% do lower the selectivity towards C_2's and increase the selectivity towards carbon monoxide.

The effect of the feed composition on the fractional conversion is shown in Figure 6. The feed composition was the only parameter changed between the two systems. The fractional methane conversion is higher for the 2% oxygen system than the 20% oxygen system. However, the methane reaction rate, not shown, is higher for the 20% oxygen system due to the higher throughput of methane in the system. The conversion in both conditions increases with an increase in residence time.

Hydrogen, acetylene, and carbon monoxide are still the major products of both systems, while carbon dioxide is still very low and water is negligible for both systems. However, the selectivity of the major products is different in the two systems. Acetylene still comprises 90% of the C2's with ethylene and ethane around 6% and 4%, respectively.

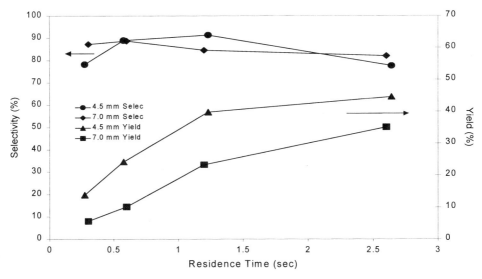

Figure 3. Hydrogen selectivity and yield for the two different reactor cross sectional areas at different residence times. Feed composition, 4/1 CH_4/O_2; gas temperature, 298 K; power, 5.2 watts.

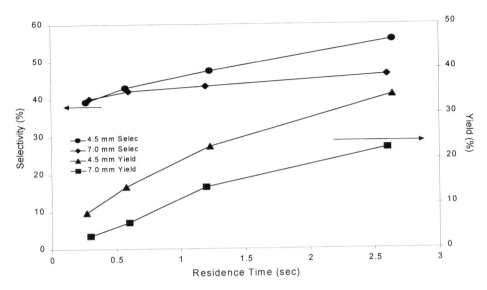

Figure 4. Acetylene selectivity and yield for the two different reactor cross sectional areas at different residence times. Feed composition, 4/1 CH_4/O_2; gas temperature, 298 K; power, 5.2 watts.

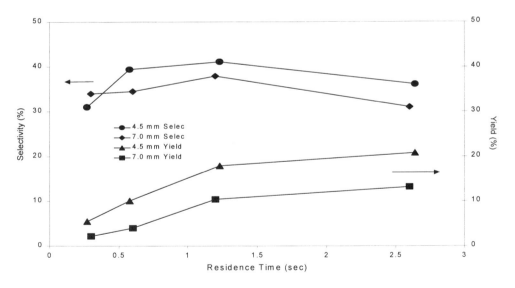

Figure 5. Carbon monoxide selectivity and yield for the two different reactor cross sectional areas at different residence times. Feed composition, 4/1 CH_4/O_2; gas temperature, 298 K; power, 5.2 watts.

63

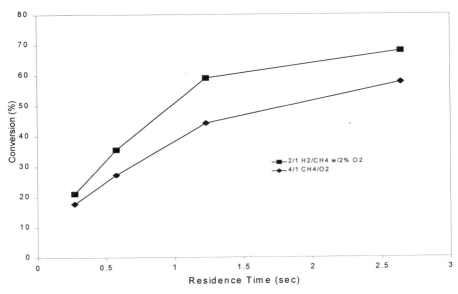

Figure 6. Comparison of methane conversion versus residence time for two different systems: feed compositions of 2/1 H_2/CH_4 with 2% oxygen, and 4/1 CH_4/O_2. Gas temperature, 298 K; power, 5.2 watts.

Figures 7-9 show the effect of residence time on the selectivity and yield of the two different feed compositions. The yield basis for hydrogen includes only the net hydrogen produced from the reacted methane. The selectivity towards hydrogen, shown in Figure 7, is lower for the system that contains hydrogen in the feed. However, the conversion is higher in that system resulting in a hydrogen yield that is essentially the same in both systems at each residence time.

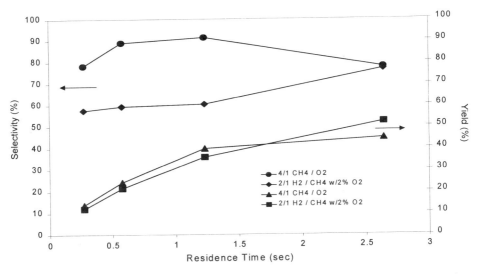

Figure 7. Hydrogen selectivity and yield for the two different feed compositions at different residence times. Feed compositions of 2/1 H_2/CH_4 with 2% oxygen or 4/1 CH_4/O_2; gas temperature, 298 K; power, 5.2 watts.

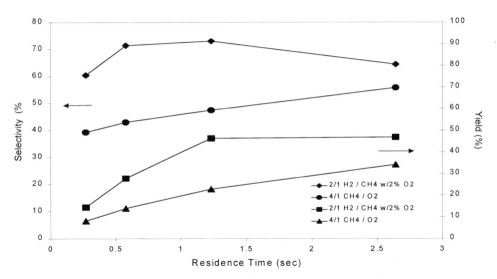

Figure 8. Acetylene selectivity and yield for the two different feed compositions at different residence times. Feed compositions of 2/1 H_2/CH_4 with 2% oxygen or 4/1 CH_4/O_2; gas temperature, 298 K; power, 5.2 watts.

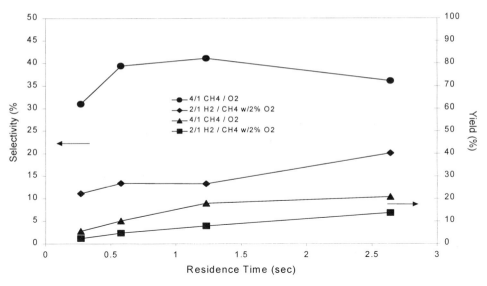

Figure 9. Carbon monoxide selectivity and yield for the two different feed compositions at different residence times. Feed compositions of 2/1 H_2/CH_4 with 2% oxygen or 4/1 CH_4/O_2; gas temperature, 298 K; power, 5.2 watts.

With the lower oxygen concentration, the 2% oxygen system shows a higher selectivity toward acetylene (Figure 8) and a lower selectivity toward carbon monoxide (Figure 9).

The high concentration of oxygen in the feed for the 20% oxygen system causes an over oxidation of the carbon species to CO, reducing the selectivity towards acetylene. This product flexibility could be useful in that it would allow for hydrogen to be produced with acetylene for higher hydrocarbons or with carbon monoxide to produce a synthesis gas feed stream. Synthesis gas production is a very costly step in many industrial processes.

The production of hydrogen from methane is important because it has a low environmental impact. The production of hydrogen via steam reforming has a H_2/CO_2 ratio of 4 (Table 1), the lowest CO_2 impact of any fossil fuel source. The dc plasma system produces a H_2/CO_2 ratio of about 1000 in the effluent stream. However, CO is a major product of the system. If the water-gas shift reaction were used to convert all of the produced CO into CO_2 via reaction [5], the resulting H_2/CO_2 ratio would be approximately 9 for either system. This value is still considerably better than that of other fossil fuels, including steam reforming.

Hydrogen is not the only valuable resource produced in this system. As previously mentioned, the other by-products are acetylene and carbon monoxide. These products can return some value from either their heating or chemical value. The product gas stream of unreacted methane, acetylene, and carbon monoxide retains over 90% of the heating value that was put into the system via methane. However, the major cost of hydrogen production in this system is the energy required to obtain the plasma discharge. Currently, the production of hydrogen from methane in the catalytic dc system costs about four times the projected goal for the cost of hydrogen. However, the cost of hydrogen could be reduced if the acetylene was sold for its chemical value or even further processed to a higher value chemical.

CONCLUSIONS

The dc plasma catalytic system is very effective in the conversion of methane to hydrogen, acetylene, and carbon monoxide. Reducing the cross sectional area of the reactor decreased the amount of gas that was bypassing the streamer discharges resulting in an increase in methane conversion. Single pass methane conversions as high as 68% and hydrogen, acetylene, and carbon monoxide yields of 52%, 47%, and 21%, respectively, have been achieved. High hydrogen yields can be achieved under different conditions. The highest conversions were obtained with an oxygen concentration of 2% and a residence time of 2.6 seconds. Further work needs to be done to reduce the energy cost. The projected cost of hydrogen may be met by increasing conversion and the throughput of methane while maintaining similar power requirements. This could be accomplished by further minimizing bypassing to increase the overall efficiency of the plasma zone.

ACKNOWLEDGMENTS

The authors would like to express their appreciation to the United States Department of Energy (Contract DE-FG21-94MC31170) and the National Science Foundation for sponsoring a Graduate Research Traineeship in Environmentally Friendly Natural Gas Technologies. Some of the experimental work was performed by Duc Le, and his help is greatly appreciated. In addition, the assistance from P. Howard, T. Caldwell, D. Larkin, H. Le, and L. Zhou is appreciated.

REFERENCES

Armor, J. N. (1999). "The multiple roles for catalysis in the production of H2." *Applied Catalysis A: General* **176**: 159-176.

Balasubramanian, B., A. Lopez Ortiz, S. Kaytakoglu, D.P. Harrison (1999). "Hydrogen from methane in a single-step process." *Chemical Engineering Science* **54**: 3543-3552.

Liu, C. J., R. G. Mallinson, L. Lobban. (1998). "Nonoxidative methane conversion to acetylene over zeolite in a low temperature plasma." *Journal of catalysis* **179**: 326-334.

Liu, C.-J., A. Marafee, B Hill, G. Xu, R.G. Mallinson, L. Lobban. (1996). "Oxidative coupling of methane with ac and dc corona discharge." *Industrial & Engineering Chemistry Research* **35**(10): 3295-3301.

Marafee, A., C.-J. Liu, G. Xu, R.G. Mallinson, L. Lobban. (1997). "An experimental study on the oxidative coupling of methane in a direct current corona discharge reactor over Sr/La2O3 catalyst." *Industrial & Engineering Chemistry Research* **36**: 632-637.

Scholz, W. H. (1993). "Processes for industrial production of hydrogen and associated environmental effects." *Gas separation & purification* **7**(3): 131.

A NOVEL CATALYTIC PROCESS FOR GENERATING HYDROGEN GAS FROM AQUEOUS BOROHYDRIDE SOLUTIONS

Steven C. Amendola, Michael Binder, Michael T. Kelly, Phillip J. Petillo, and Stefanie L. Sharp-Goldman

MILLENNIUM CELL LLC
1 Industrial Way West
Eatontown, NJ 07724

ABSTRACT

A safe, simple, lightweight, and compact process generates high-purity hydrogen gas on demand from base-stabilized, aqueous solutions of sodium borohydride, $NaBH_4$, by using a ruthenium catalyst. $NaBH_4$ solutions can generate the equivalent of >2,500 Wh/L and >7% H_2 by weight can be recovered. These inherently stable solutions do not generate significant amounts of H_2 gas under ambient conditions. However, when in contact with heterogeneous Ru catalyst, $NaBH_4$ solutions rapidly hydrolyze to form H_2 gas and sodium borate, a water-soluble, inert salt. H_2 generation only occurs when $NaBH_4$ solutions are in contact with Ru catalyst. When Ru catalyst is removed from $NaBH_4$ solution (or $NaBH_4$ solution is separated from Ru catalyst), H_2 generation stops. This H_2 generator promises to be safer, have quicker response to H_2 demand, have a greater H_2 storage efficiency, and be more easily controllable than current H_2 storage devices/generators. It can be easily incorporated into any system where H_2 gas is required, such as powering internal combustion engines or fuel cells.

INTRODUCTION

Hydrogen gas is clearly one of the best alternatives to fossil fuels since it has higher energy efficiencies and lower emissions than can be attained with hydrocarbon fuels. H_2 gas is an environmentally desirable fuel for transportation applications since it can be either used directly as a combustion fuel in internal combustion engines or electrochemically oxidized in a variety of fuel cells. In both cases, in addition to energy, water is the primary reaction by-product. Although H_2 is extremely energy rich on a weight basis, it is relatively poor on a volumetric basis. Because H_2 gas has such a low density, large

Advances in Hydrogen Energy, edited by Padró and Lau
Kluwer Academic/Plenum Publishers, 2000

volumes of H_2 must be safely generated or stored onboard a vehicle to enable significant range. The low volumetric efficiency of various H_2 storage systems has precluded its use in practical vehicles.

A new, uncomplicated approach for production, transmission, and storage of H_2 has been developed. This H_2 generation system is based on an aqueous solution that acts as the H_2 carrier and storage medium. When H_2 is needed at the point of use, a catalytic reaction is used to generate high-purity H_2 gas from the solution. Using a liquid to deliver/generate H_2 permits vehicle refueling similar to current gasoline stations, and the possibility of achieving safe, practical H_2 powered vehicles is enhanced.

BOROHYDRIDE H_2 GENERATOR

In this section, details of an easily controllable, safe method for producing high-purity H_2 gas are described. This method of generating H_2 gas is particularly suitable for providing a clean source of H_2 gas for use as an anodic fuel in fuel cells or as a fuel for internal combustion engines in transportation applications. This compact, portable H_2 generator is based on a non-pressurized, aqueous solution of alkaline sodium borohydride ($NaBH_4$, tetrahydroborate). As found by Schlesinger et al.,[1] when aqueous $NaBH_4$ solutions contact selected metal (or metal boride) catalysts, these solutions hydrolyze to yield H_2 gas and water-soluble, sodium borate. Overall reaction stoichiometry can be represented in a simplified form as:

$$NaBH_4(aq) + 2 H_2O \xrightarrow{\text{catalyst}} 4 H_2 + NaBO_2(aq) \qquad [1]$$

Reaction [1] would occur to some extent even without a catalyst if the solution pH < 9. However, to increase shelf life of $NaBH_4$ solutions (and to prevent H_2 gas from being slowly produced upon standing), $NaBH_4$ solutions are typically maintained as a strongly alkaline solution by adding NaOH. The key feature of using Reaction [1] to produce H_2 is that H_2 generation in alkaline $NaBH_4$ solutions (>pH 14) occurs only when these solutions are allowed to contact selected heterogeneous catalysts. Without catalysts present, strongly alkaline $NaBH_4$ solutions do not produce appreciable H_2. This ensures a rapid, dependable, and controllable response to H_2 demand because H_2 is generated only when selected catalysts contact $NaBH_4$ solution. When Ru catalyst is separated from the alkaline $NaBH_4$ solution, H_2 generation stops. No addition of water, acid, pressure, or heat is required to generate H_2. Since four moles of H_2 gas are catalytically generated for each mole of $NaBH_4$ reacted, $NaBH_4$ solutions act as a means of compactly storing H_2 gas. Reaction [1] is extremely efficient on a weight basis since one half of the H_2 produced is obtained from H_2O and one from $NaBH_4$.

Since base-stabilized $NaBH_4$ solutions not in contact with catalyst are stable and produce virtually no H_2, in the event of a leak or a spill, no H_2 will be generated. Because little free H_2 is actually stored in the system, concerns about onboard bulk H_2 storage or distribution are reduced. As $NaBH_4$ solutions are easier to store onboard a vehicle than H_2 gas, these solutions offer a practical alternative to direct H_2 fueling. $NaBH_4$ solutions are also easier and safer to distribute to consumers than bottled H_2 gas and can be easily transported from terminal locations to service stations via truck. $NaBH_4$ solutions, with a viscosity and density close to that of water, can be dispensed in the same manner as gasoline with minor modification to the dispensing equipment.

The other hydrolysis product of Reaction [1], sodium borate, is water soluble and environmentally innocuous. Borates have not been shown to pose environmental hazards in water supplies. Since Reaction [1] is totally inorganic, it produces no fuel cell poisons (no sulfur, soot, CO, or aromatics are formed). Only H_2 and some water vapor are produced in the product gas stream. This is beneficial for use in proton exchange membrane (PEM) fuel cells, where the water vapor can be used to humidify the membranes. The borate reaction product can be utilized as a starting material in regenerating $NaBH_4$.

Aqueous, alkaline $NaBH_4$ solutions are a safe method for generating H_2 since these non-flammable solutions are stable in air. These properties are unique among metal hydride systems that generate H_2 by reaction with water. The heat generated by Reaction [1] (~70 kJ/mole H_2 formed), is considerably less than the typical >125 kJ/mole H_2 produced by reacting other chemical hydrides with water.[2] To illustrate differences in reactivities, if LiH, NaH, $LiAlH_4$, or other alkali/alkaline-earth hydride is added to water in the presence of air, they spontaneously ignite. When $NaBH_4$ is added to water some H_2 will be evolved initially, and the reaction rate falls while H_2 is formed with slow bubbling. If the water is sufficiently basic, little or no H_2 generation occurs. It should be noted that lithium, potassium or ammonium borohydride solutions can also be used as H_2 sources. $NaBH_4$, however, has greater water solubility than other borohydride salts so it is convenient and simpler to handle. For this reason, $NaBH_4$ solutions were studied.

To summarize, generating H_2 catalytically via Reaction [1] has the following advantages that will make H_2 powered transportation a reality:

- $NaBH_4$ solutions are nonflammable
- $NaBH_4$ solutions are stable in air for months
- H_2 generation only occurs in the presence of selected catalysts
- The only other product in the H_2 gas stream is water vapor
- Reaction products are environmentally safe
- H_2 generation rates are easily controlled
- Volumetric and gravimetric H_2 storage efficiencies are high
- The reaction products can be recycled
- H_2 can be generated even at $0^\circ C$.

SODIUM BOROHYDRIDE

Sodium borohydride solutions are a convenient, compact way to store H_2. For a discussion of the theoretical amount of H_2 available to a $NaBH_4$ solution, please see Appendix A. It is more effective to store energy in the form of chemical bonds (within $NaBH_4$) than it is to store energy as pressure, as evidenced by the low volumetric efficiencies of pressurized H_2 gas. The most critical aspect of using $NaBH_4$ solutions as a means of H_2 storage is the quantity of H_2 that is contained in a given weight and volume of solution.

The solubility limit of $NaBH_4$ in water is about 35 wt% at 25°C. Assuming 100% stoichiometric yield in Reaction [1], one liter of 35 wt% $NaBH_4$, 5 wt% NaOH aqueous solution (measured density = 1.05 g/mL) yields:

$$(1000 \text{ mL}) \text{ x } (1.05 \text{ g/mL}) \text{ x } (0.35\text{g } NaBH_4/\text{g sol'n}) / (37.83 \text{ g/mole}) = 9.7 \text{ mol } NaBH_4 \quad [2]$$

$$(9.7 \text{ moles } NaBH_4) \text{ x } (4 \text{ mole } H_2/\text{mole } NaBH_4) \text{ x } (2 \text{ g } H_2/\text{mole } H_2) = 78 \text{ grams } H_2 \quad [3]$$

$$78 \text{ g } H_2 / 1050 \text{ g } NaBH_4 \text{ solution} = \underline{7.4\% \text{ hydrogen by weight}} \quad [4]$$

Functional prototype H_2 generators have been built that generated H_2 at rates >3 L/min and yielded >95% of the theoretical H_2 available from a 30 wt % solution of $NaBH_4$ (to yield 6.3% H_2 by weight).

With proper engineering, and assuming that 90% of the water generated by either a fuel cell or H_2 burning engine can be recaptured and returned to the H_2 generator, it is probable that borate solubility problems can be minimized. This would allow fully saturated $NaBH_4$ solutions (ie. slurries) to be employed. This is an important factor since, if at a given time and temperature, an insufficient amount of water remains in the system to fully solvate the reaction product, sodium borate will precipitate and clog the system.

The following summarizes the results of the sample calculations to illustrate the amounts of H_2 available from $NaBH_4$ solutions. The measured density of 35 wt % $NaBH_4$ solutions is 1.05 g/cm^3. This low value for the solution is not totally surprising since the density of $NaBH_4$ powder is 1.07g/cm^3.

$$2H_2O + NaBH_4 \rightarrow 4 \text{ moles } H_2$$
$$\rightarrow 8 \text{ g } H_2 \text{ or } 98 \text{ liters at } 25^\circ C \qquad [5]$$
$$1 \text{ liter of 35 wt \% } NaBH_4 \text{ solution} \rightarrow 960 \text{ liters } H_2 \text{ or } 78 \text{ g } H_2$$

Storing H_2 in aqueous $NaBH_4$ solutions compares favorably to other H_2 storage methods. For example, the volume required to store 5 kg of liquid H_2 in cryogenic containers is ~110 liters. In pressurized tanks, storing 5 kg H_2 at 34 MPa (5000 psi) requires ~ 220 liters. However, storing 5 kg of H_2 in 35 wt % $NaBH_4$ solutions requires only 65 liters.

H_2 PRODUCTION

Although Equation [5] suggests that aqueous $NaBH_4$ solutions can store sufficient H_2 to make a practical fuel, the rate at which H_2 can be extracted or produced is very important. Experiments with $NaBH_4$ solutions have been conducted in two modes: the static $NaBH_4$ solution and the flowing $NaBH_4$ solution.

In a static system, catalyst was placed in a stainless steel basket and dropped into a solution containing various concentrations of $NaBH_4$. The H_2 gas generated was collected in inverted graduated cylinders and the volume of H_2 gas per unit time was measured. The H_2 generation rate via Reaction [1] depends on the type of catalyst used, its active surface area, solution concentrations of $NaBH_4$ and NaOH, and temperature. Figure 1 shows a typical plot of initial H_2 volumes generated by ruthenium catalyst supported on ion exchange resin beads of $NaBH_4$ solutions as a function of time. H_2 volumes generated increased linearly with time. This relationship can be expressed as:

$$d[H_2]/dt = K \bullet W \qquad [6]$$

Here, the rate at which H_2 gas is generated from $NaBH_4$ solutions as a function of time, $d[H_2]/dt$, in units of liters H_2 generated per second, is the product of K, the rate constant for catalytic hydrolysis (which depends on the type of catalyst support, $NaBH_4$ and NaOH concentrations, and solution temperature), and W, the weight of supported Ru catalyst (which is directly related to metal catalyst surface area). The amount of H_2 that can be generated per second from a given $NaBH_4$ solution may be increased by simply increasing the amount of catalyst, i.e. increasing the term W in Equation [6].

The fact that a plot of H_2 volumes initially generated vs. time gave a straight line is indicative of pseudo zero order kinetics. For borohydride hydrolysis, Kaufman and Sen[3] and Holbrook and Twist[4] also found zero order kinetics. Zero order kinetics for Reaction [1] imply that hydrolysis is independent of the concentrations of the reacting chemical species. This can be explained by assuming that the initial reaction step probably involves a surface catalyzed reaction, most likely BH_4^- adsorption on the catalyst. Since the number of active sites on the catalyst surface is constant (that is, catalyst is not consumed or deactivated by reaction products), the volume of H_2 generated versus time appears as a straight line. As Reaction [1] proceeds and $NaBH_4$ is depleted, $NaBH_4$ concentration eventually becomes the limiting factor and zero order kinetics are no longer observed. This is shown in Figure 2, where a plot of generated H_2 vs. time for low starting $NaBH_4$ concentration deviates from linearity at longer times. It has also been observed that water concentration is important.

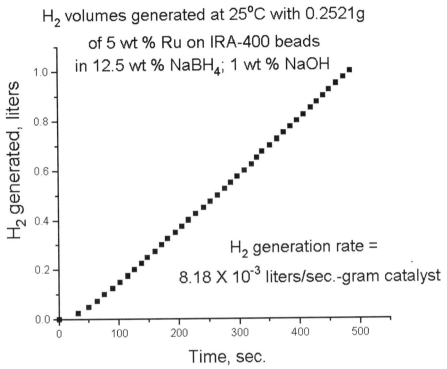

Figure 1. H_2 volume generated as a function of time with 0.25g of 5 wt % Ru supported on IRA 400 anion exchange resin in 12.5 wt % $NaBH_4$, 1 wt % NaOH, 86.5 wt % solution.

Initial H_2 generation rates as a function of wt% $NaBH_4$, (for wt% NaOH held constant at 1, 5, or 10 wt%) and constant catalyst loading are shown in Figure 3. As the wt% $NaBH_4$ increases, the initial H_2 generation rate increases and reaches a maximum in the range of 7.5 - 12.5 wt% $NaBH_4$. At higher wt% $NaBH_4$, the initial H_2 generation rate decreases. $NaBH_4$ solution compositions having the highest initial H_2 generation rates (e.g. 1 wt% NaOH and ~12.5 wt% $NaBH_4$) are not necessarily ideal for overall H_2 capacity (total amount of H_2 that can be delivered) or for long term shelf life. H_2 generation rates shown in Figure 3 are, however, already sufficient for high fuel demand (high power) applications such as vehicle acceleration.

The second general trend shown in Figure 3 was that as wt% NaOH decreases from 10 wt% to 1 wt%, H_2 generation rates for a given wt% $NaBH_4$ markedly increase. A 7.5 wt% $NaBH_4$ solution not containing any added NaOH catalyzed by 5 wt% Ru at 25°C has a H_2 generation rate of 7.75 milliliters/sec-gram catalyst. Without adding NaOH, solution pH rises to about 9, but the solution still exhibits slow hydrolysis even without a catalyst present. Clearly, a small amount of NaOH must be present in $NaBH_4$ solutions to extend solution shelf life.

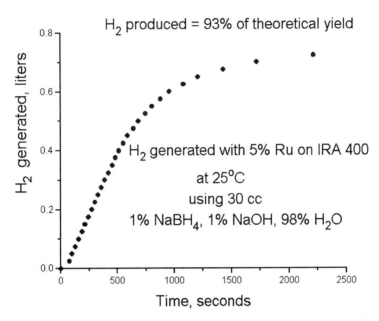

Figure 2. H$_2$ volume generated as a function of time with ~ 0.25g of 5 wt % Ru supported on IRA 400 anion exchange resin in 1 wt % NaBH$_4$, 1 wt % NaOH, 98 wt% H$_2$O solution.

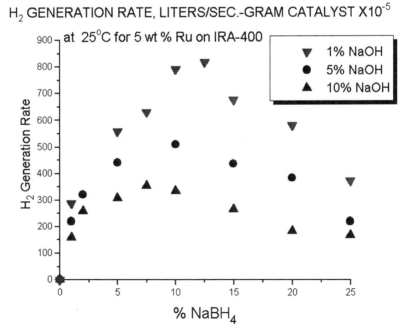

Figure 3. H$_2$ generation rates with 5 wt % Ru supported on IRA 400 anion exchange resin in various wt % NaBH$_4$. NaOH was held constant at 10, 5, or 1 wt %.

H_2 generation rates were measured in 7.5 wt% $NaBH_4$, 1 wt% NaOH, and 91.5 wt% water solutions at 0°C, 25°C, 32.5°C, and 40°C. It is noteworthy that H_2 could be generated from $NaBH_4$ solutions even at 0°C. Measured H_2 generation rates were:

- 0.76 milliliters/sec-gram catalyst at 0°C
- 6.3 milliliters/sec-gram catalyst at 25°C
- 10.1 milliliters/sec-gram catalyst at 32.5°C
- 18.3 milliliters/sec-gram catalyst at 40°C.

An Arrhenius plot of log [initial H_2 generation rate] vs. reciprocal absolute temperature (1/T) is shown in Figure 4. The slope of the straight line yields an apparent Ru catalyzed activation energy of 56 kJ mole[-1] (r^2 = 0.999) for the 0-40°C temperature region of interest. This value compares favorably with activation energies found by Kaufman and Sen[3] for $NaBH_4$ hydrolyzed with other metals (Co: 75 kJ mole[-1], Ni: 71 kJ mole[-1], and Raney Ni: 63 kJ mole[-1]). Under normal operating conditions, as the exothermic H_2 generation Reaction [1] proceeds, the heat generated raises $NaBH_4$ solution temperatures, which tends to further increase H_2 generation rates.

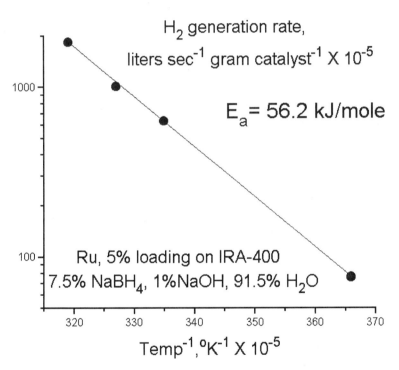

Figure 4. Log [H_2 generation rates] plotted against 1/T. The solution, 7.5 wt% $NaBH_4$, 1 wt% NaOH, 91.5 wt% H_2O, was catalyzed by ~0.25g of 5 wt% Ru supported on IRA 400 anion exchange resin at 40°C, 32.5°C, 25°C and 0°C. The apparent activation energy calculated from the slope of the straight line was 56 kJ mole[-1].

Until now, we have discussed H_2 generation via supported catalyst immersed in quiescent $NaBH_4$ solution - no flowing/stirring solutions. Here, diffusion and convection are the main pathways for bringing fresh $NaBH_4$ to catalyst surfaces and for removing formed H_2 gas from catalyst surfaces. These slow processes limit H_2 generation rates. In addition, when the sodium borate solubility limit in solution is exceeded (by loss of water and formation of sodium borate reaction product), sodium borate may precipitate near/on the catalyst surface thereby reducing subsequent catalyst activity and H_2 generation efficiency.

To eliminate these drawbacks and ensure that catalysts are adequately supplied with $NaBH_4$ while H_2 gas and sodium borate is continually removed, a flow system was designed where circulating $NaBH_4$ solution passes through a container enclosing supported catalyst. Advantages of a flow system over a quiescent system include:

- fast H_2 generation rates
- better thermal management
- superior water management
- greater (>95%) H_2 yields due to less precipitation of borates on the catalyst.

Figure 5 shows typical H_2 volumes generated by a flow system. Two important points are immediately evident from Figure 5. Firstly, H_2 generation is extremely rapid, with an average rate of 42.3 milliliters/sec-gram catalyst for the first 90% of the available H_2, with a maximum rate of 111 milliliters/sec-gram. Secondly, overall H_2 yields based on stoichiometric amounts of $NaBH_4$ is >95%. H_2 generation rates decrease as the solution becomes more slurry-like and less water is available for reaction.

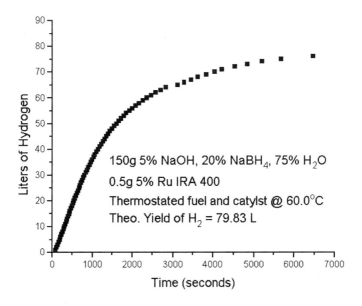

Figure 5. H_2 generation rates with 5 wt% Ru supported on IRA 400 anion exchange resin at 60°C in a flowing system. The flow rate was 112 ml/min.

THE CATALYST

Levy et al.[5] and Kaufman and Sen[3] investigated cobalt and nickel borides as catalysts for controlled generation of H_2 from $NaBH_4$ solutions. Brown and Brown[6] investigated various metal salts and found that ruthenium and rhodium salts liberated H_2 most rapidly from borohydride solutions. These results, along with our own work, indicated that ruthenium would most likely be the most active metal for catalysis of

Reaction [1]. Work finding a suitable substrate for making a practical H_2 generator therefore started with ruthenium to verify if supported Ru would maintain the highest activity among the candidate materials. Many substrate systems were tried but the mere presence of a certain percentage of Ru did not guarantee rapid hydrolysis rates. This indicates that the nature of the substrate and preparation technique is very important. Ion exchange resins proved to be relatively stable in alkaline $NaBH_4$ solutions. Nevertheless, even with ion exchange resin supports, vigorous evolution of H_2 bubbles can destroy the physical integrity of the catalyst layer. Tiny Ru particles floating on the solution after generating liters of H_2 were often observed in static systems. In flowing systems the ability to use finer mesh screens eliminates this problem.

Both anionic and cationic exchange resin beads were analyzed as catalyst supports for Ru in activating borohydride hydrolysis. The criteria used for evaluation was the initial H_2 generation rate constant, K, in units of milliliters H_2 generated per second per gram of supported Ru catalyst. In Table 1, K values for various ion exchange resin bead supports are reported. For catalytically generating H_2 from $NaBH_4$ solutions, anionic resins are considerably better catalyst supports for Ru than cationic resins since anionic resins have discernibly higher H_2 production rates. Specifically, A-26 and IRA-400 (Rohm and Haas) anion exchange resins, when used as Ru catalyst supports, gave the fastest H_2 generation rates. H_2 generation rates for A-26 and IRA-400 anion exchange resins varied linearly with catalyst weights (doubling the catalyst weight approximately doubles H_2 generation rates). Since catalyst weight is directly related to overall catalyst surface areas, $NaBH_4$ hydrolysis (and H_2 generation) is related to and rate limited by Ru content.

Table 1. Rate of H_2 gas generated catalytically per gram of catalyst (Ru + support resin) are compared for various anionic and cationic support resins in 20% $NaBH_4$, 10% NaOH, 70% H_2O solution at 25°C. All catalysts had ~5 wt% Ru loading.

Ru Resin Support	Time to generate 1000 milliliter H_2 gas (seconds)	Weight of Ru + support (grams)	Milliliters H_2/sec per gram catalyst
Anionic Resins			
A-26	1161	0.2563	3.36
A-26	633	0.5039	3.13
IRA-400	1173	0.2565	3.32
IRA-400	773	0.4116	3.14
IRA-900	1983	0.2555	1.97
Dowex 550A	672	0.7692	1.93
Dowex MSA-1	791	0.7691	1.64
Dowex MSA-2	1028	0.7691	1.26
A-36	1415	0.2550	1.11
Cationic Resins			
MSC-1B	2351	0.2592	1.64
Dowex HCR-W2	895	0.7631	1.46
MSC-1A	1382	0.5054	1.43
Amberlyst 15	2871	0.2563	1.36
Amberlyst 15	1324	0.5054	1.49
Dowex 22	1818	0.7678	0.72
Dowex 88	6163	0.2556	0.63

Ion exchange resin beads not coated with Ru did not show any catalytic activity for generating H_2 from $NaBH_4$ solutions. Differences in catalytic activity among different types of resin beads probably depend on interactions between BH_4^- anions and active sites on anionic exchange resins. This hypothesis was tested by baking a sample of IRA-400 anionic exchange resin in a vacuum furnace for 15 hours at 100°C. At temperatures above 60°C, amine groups are driven off the resin. After this treatment, catalyst activity decreased significantly. Thus, it is likely that there is some interaction between the positively charged amines present on the resin and the negatively charged borohydride ions in solution. This speculation is supported by the fact that when anionic resins, containing amine groups, are used as Ru supports, they produce H_2 at faster rates than cationic resins, which do not contain amine groups. In our flow system, we have generated over 1500 L of H_2 per gram of Ru catalyst with no observed degradation in catalyst performance. This suggests that long Ru catalyst lifetimes can be expected.

SYSTEM POWER ANALYSIS

The data from Figure 5 indicate an initially high rate of H_2 generation in a flowing system slows down towards the end of the reaction. If the average rate of generation over the duration of the experiment is taken it is found that 1 liter H_2 can be generated in 47 seconds using only 0.5 g Ru catalyst from a solution containing 20 wt% $NaBH_4$ and 5 wt% NaOH. This rate of H_2 production, the equivalent of ~100 watts, was obtained with only 0.5 g of catalyst, of which 5% is Ru. A more detailed discussion of how hydrogen generation rate can be converted into power is given in Appendix B. This power level is the equivalent of ~ 4 kW/g of active Ru at 60°C. Faster H_2 generation rates (and power levels) may be expected for more efficient catalysts, catalysts with better dispersion on the substrate, or substrates with higher Ru loading.

According to US Department of Energy (DOE) sponsored studies, it is estimated that powering a car requires ~ 0.3 kWh/mile.[7] Since 1 kWh requires 741 liters H_2 at 50% efficiency (see Appendix B), traveling 1 mile (0.3 kWh) requires 20 g H_2. Since this amount of H_2 can be produced by 0.26 liters of a 35 wt % $NaBH_4$ solution, these $NaBH_4$ solutions have an effective mileage equivalent of ~14.5 miles per gallon. This is comparable to what is obtainable with fossil fuels. A car traveling at 60 mph (1 mile/min) will use 0.3 kWh in 1 minute. This is a power requirement of 18 kW which can be achieved by utilizing only 4.5 g of active Ru (18 kW / 4 kW/g Ru) or 90 g of supported Ru catalyst.

PROTOTYPE H_2 GENERATORS

Numerous designs for portable H_2 generators using catalysis of $NaBH_4$ solutions have been successfully built and tested. One design, shown schematically in Figure 6, has no moving parts and uses pressure equalization to move $NaBH_4$ solutions from a reservoir into contact with a tube made out of stainless steel screen, which contains catalyst. While H_2 is being generated, pressure differences force $NaBH_4$ solution away from the catalyst tube to stop H_2 generation. When sufficient H_2 has been consumed, the pressure drop will once again allow $NaBH_4$ solution to contact the catalyst. This design is compact, self-regulating, and requires no external power source.

Another design, illustrated in Figure 7, uses a mechanical pump to meter $NaBH_4$ solution through a catalyst bed to generate H_2. The catalyst chamber contains a stainless steel screen and the solution contacts catalyst through the screen. This design has very rapid response to H_2 requirement. The pump runs $NaBH_4$ solution over the catalyst and pressure builds up. Above a certain pressure, the pump reverses direction to drain the catalyst chamber and thereby stops H_2 production. The pump then shuts off. When H_2 pressures drop below a certain predetermined point, the pump turns back on in the forward direction to refill the catalyst chamber, generate additional H_2 and replenish H_2 pressures.

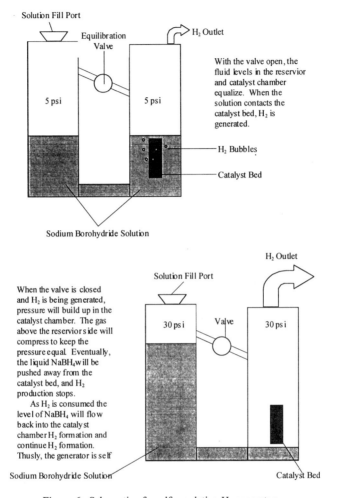

Figure 6. Schematic of a self-regulating H_2 generator.

Figure 8 is a photograph of a prototype series hybrid engine. It includes a H_2 generator of the type shown in Figure 7, a 3 kW internal combustion engine that has been converted from running on gasoline to running on H_2, and a 1 kW alternator. A (discharged) standard automotive battery, electric motor, wheel drive, and lights serve as a load. Depending on the battery state of charge, running this engine requires between 5 and 12 liters of H_2 per minute.

Figure 9 shows a close up view of the reservoir. A regulator and safety valve fitted to the top indicate pressure inside the reservoir. When sufficient H_2 pressure builds up inside the reservoir, the pump can be reversed to drain the catalyst chamber. With no solution is in the catalyst chamber, no H_2 is generated. The catalyst chamber shown in Figure 10 is 3 inches long and 3/4 inches in diameter and contains enough catalyst (less than 5 grams) to supply the 5 to 12 liters per minute H_2 required to run the engine.

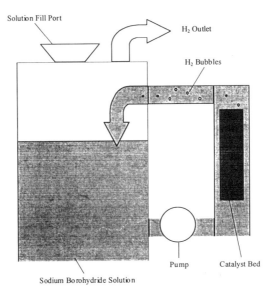

The NaBH$_4$ is pumped to the catalyst chamber, where it makes contact with the catalyst, generates H$_2$, and returns to the main chamber. When sufficient H$_2$ has been generated and additional H$_2$ is no longer needed, an external sensor triggers the pump to reverse direction and drains the catalyst chamber, thereby ceasing H$_2$ production.

Figure 7. Schematic of a H$_2$ generator regulated by a mechanical pump.

Figure 8. Prototype of a H$_2$ burning internal combustion engine/electric motor hybrid. On the lower left is the H$_2$ generator. On the lower right is the mechanical metering pump and electric motor. The upper right is the 3 kW engine, upper left is a standard automotive battery that can be charged by an alternator.

Figure 9. Close-up view of the reservoir portion of the H$_2$ generator shown in Figure 8. Also shown are standard automotive headlights that draw power from the alternator.

Figure 10. The catalyst chamber of the H$_2$ generator shown in Figure 8. The chamber is 3 inches long and ¾ inches in diameter. It contains less than 5 grams of catalyst, enough to supply sufficient H$_2$ to run the engine.

COST OF NaBH₄

The combined advantages of using NaBH$_4$ solutions as a H$_2$ source are so compelling as to provide an economic stimulus to produce NaBH$_4$ as a commodity chemical. This would lower its current cost, \$40/kg, to the point where it would be economically competitive with fossil fuels and therefore feasible for use in transportation. The current cost of NaBH$_4$ is due to the small scale of manufacture. To reach full commercialization of this promising technology, and ensure its use in new applications, large scale manufacturing methods and/or new synthesis will produce the required lower production costs. Implicit in this method is to use the borate reaction product as the starting material for NaBH$_4$ synthesis. In this way, NaBH$_4$ refueling, recharging, and recycling facilities could assume parallel roles in a NaBH$_4$-based transportation systems to that held by oil refineries in today's fuel distribution system, but without the negative environmental impacts of refineries. Current pipelines and service stations can be leveraged to transport NaBH$_4$ solutions rather than gasoline, at a cost savings to consumers. This change could be accomplished by retrofitting existing infrastructures.

EXPERIMENTAL

Catalyst Preparation

Ru supported/dispersed on ion exchange resin beads was prepared by a two-step process (incipient wetness method). The first step involved ion exchange. The procedure used for supporting ruthenium on ion exchange resin supports was dependent on whether the resin was cationic or anionic in nature. For cationic resins, an amount of RuCl$_3$ (Alfa-Aesar) necessary for a 5 wt% Ru loading was dissolved in deionized water. The RuCl$_3$ solution was added to a weighed amount of cation ion exchange resin beads that had been previously washed and dried at 50°C. The slurry of RuCl$_3$ solutions and ion exchange resin was allowed to stand at ambient temperature for 24 hours with stirring at regular intervals to maintain uniformity. During this impregnation stage, [Ru]$^{+3}$ would be adsorbed on the negative sites of the ion exchange resin due to charge interaction. The slurry was then dried by evaporation at 50°C, to allow any excess RuCl$_3$ to adsorb onto the porous material.

To prepare Ru supported on anion exchange resins appropriate amounts of RuCl$_3$-3H$_2$O were dissolved in deionized water and acidified with HCl to convert RuCl$_3$ into [RuCl$_6$]$^{-3}$. This H$_3$RuCl$_6$ solution was added to a weighed amount of anion exchange resin beads that had been previously washed and dried at 50°C. The slurry of H$_3$RuCl$_6$ solution and ion exchange resin beads was allowed to stand at ambient temperature for 24 hours with stirring at regular intervals to maintain uniformity. During this impregnation stage, [RuCl$_6$]$^{-3}$ would be mainly adsorbed on the external positive sites of the ion exchange resins due to charge interaction. The slurry was then dried by evaporation at 50°C.

The second step in Ru catalyst preparation, which is the same for both cation and anion exchange resins, was to chemically reduce Ru ion species, now impregnated in the ion exchange resin by adding 15 wt % NaBH$_4$, 3 wt % NaOH, 82 wt % H$_2$O solution. Elemental analysis showed that the black catalyst, as prepared, contained only metallic Ru (since no boron was detected by elemental analysis, it is assumed that no borides were present). The porous nature and surface area of the resin as well as the strong interaction between Ru ions and resin prevented Ru particle aggregation.

Hydrogen Generation

In typical H_2 generation experiments reported here, ~30 ml of aqueous solution containing 1-25 wt % $NaBH_4$, and 1-10 wt % NaOH, were thermostated in a sealed flask fitted with an outlet tube for collecting evolved H_2 gas. Approximately 0.25g of supported Ru catalyst (~ 5 wt % of the 0.25 g total catalyst weight was Ru; the remainder was inert ion exchange resin support) was placed in a stainless steel screen container and dropped into the various thermostated $NaBH_4$ solutions to begin H_2 generation. $NaBH_4$ solution could contact Ru catalyst through the screen, H_2 could exit, while the lightweight, Ru-coated, resin beads, trapped inside the screen container, were prevented from floating to the top of the solution. The outlet exhaust tube from the reaction vessel was placed beneath an inverted, water filled, graduated cylinder situated in a water-filled tank. Generated H_2 was measured by monitoring water displaced from the graduated cylinder as Reaction [1] proceeded and H_2 gas was generated. Rate data was based on the times needed to generate at least 1 liter of H_2 gas.

In typical flow experiments, a solution containing 30g of $NaBH_4$ and sufficient amounts of water and NaOH to make the desired solution compositions (10-30 wt% $NaBH_4$, 5-20 wt% NaOH) was thermostated at 60°C. A peristaltic pump delivered this solution from the thermostated reservoir to a catalyst chamber (also thermostated at 60°C) containing 0.5 grams of 5 wt% Ru on IRA-400 resin. $NaBH_4$ solution was then pumped back to the reservoir completing the recirculation loop. Generated H_2 gas was allowed to escape through the top of the storage tank and collected in inverted graduated cylinders. Reactions were allowed to proceed essentially to completion (ie. until H_2 generation rates decreased considerably). Time elapsed and H_2 volumes generated were recorded.

CONCLUSIONS

Stabilized, aqueous $NaBH_4$ solutions hold promise as a safe, dependable, and effective source of high purity H_2. $NaBH_4$ solution simply contacts a catalyst to produce H_2. By using $NaBH_4$ solutions together with supported Ru catalyst, this on-demand H_2 source reduces the inherent safety concerns associated with storage and distribution of significant amounts of H_2, is more advantageous than using pressurized tanks or reactive chemical hydrides, and avoids the technical risks of on-board fuel reformers. H_2 production occurs without any heat input at ambient temperatures, and reaction products are not toxic. This system can be used in numerous applications where H_2 gas is currently used, e.g. fuel cells or H_2-fueled internal combustion engines. $NaBH_4$ generators can be quickly refueled by simply adding fresh $NaBH_4$ solution, and Ru catalysts are reusable.

ACKNOWLEDGEMENTS

We would like to acknowledge the significant contributions of Dr. M. Saleem Janjua, Nicole C. Spencer, Steven C. Petillo, and Robert Lombardo in collecting data and designing prototype generators. We have greatly benefited from numerous scientific discussions with and enlightening technical suggestions from Mike Strizki of the NJ Department of Transportation, who motivated us to initiate this study. Heartfelt kudos to Martin M. Pollak Esq. and Jerome I. Feldman Esq. of GP Strategies Inc. Their unfailing enthusiasm, constant encouragement, and optimism was the driving force throughout this effort. Our group is most grateful to Andersen-Weinroth Inc. for providing generous financial support for this applied research program. This research was partially supported by a grant from the State of New Jersey Commission on Science and Technology.

REFERENCES

1. H.I.Schlesinger, H.C.Brown, A.E.Finholt, J.R.Gilbreath, H.R.Hockstra and E.K. Hyde, "Sodium Borohydride, its Hydrolysis and its Use as a Reducing agent and in the Generation of Hydrogen", Journal of the American Chemical Society, 75, 215 (1953).

2. W.D. Davis, L.S. Mason, and G. Stegeman, "The Heats of Formation of Sodium Borohydride, Lithium Borohydride and Lithium Aluminum Hydride", J. Am. Chem. Soc. 71, 2775 (1949).

3. C.M. Kaufman and B. Sen, "Hydrogen Generation by Hydrolysis of Sodium Tetrahydroborate: Effects of Acids and Transition Metals and Their Salts", Journal of the Chemical Society, Dalton Trans. 307 (1985).

4. K.A. Holbrook and P.J. Twist, "Hydrolysis of the Borohydride Ion Catalysed by Metal-Boron Alloys", Journal of the Chemical Society (A),1971, 890.

5. A. Levy, J.B. Brown, and C.J. Lyons, "Catalyzed Hydrolysis of Sodium Borohydride", Industrial and Engineering Chemistry 52, 211(1960).

6. H.C. Brown and C.A. Brown, "New, Highly Active Metal Catalysts for the Hydrolysis of Borohydride", Journal of the American Chemical Society, 84, 1493 (1962).

7. Proceedings of the 1997 U.S. DOE Hydrogen Program Review, Business Technology Books, PO Box 574, Orinda, CA 94563.

APPENDIX A

Reactions [A1] and [A2] show the steps of the chemical process used to extract energy from sodium borohydride. [A1] is the simple hydrolysis of sodium borohydride to form hydrogen gas; [A2] is the oxidation of that hydrogen gas, either in a fuel cell or in an internal combustion engine. Reaction [A3] is the overall process, and is equal to the sum of Reactions [A1] and [A2].

$$NaBH_4 \text{ (aq)} + 2 H_2O \rightarrow 4 H_2 + NaBO_2 \text{ (aq)} \qquad \text{[A1]}$$

$$4 H_2 + 2 O_2 \rightarrow 4 H_2O \qquad \text{[A2]}$$

$$NaBH_4 \text{ (aq)} + 2 O_2 \rightarrow NaBO_2 \text{ (aq)} + 2 H_2O \qquad \text{[A3]}$$

These three reactions can be used to calculate a *theoretical yield of H2 processed per gram of sodium borohydride consumed.* This calculation assumes that all necessary water generated in [A2] can be recycled for use in [A1]. Note that Reaction [A3] indicates that the process is a net producer of water, indicating borohydride is the only limiting factor. In a perfect recycle loop, no water would ever need to be added.

Therefore, for every mole of sodium borohydride (37.83 grams) supplied to the system, four moles of H_2 (8 grams) gas are processed.

$$8 \text{ g } H_2 \text{ / } 37.83 \text{ g NaBH}_4 = 21.1 \text{ \% Hydrogen by weight} \qquad \text{[A4]}$$

Equation [A4] is a rather remarkable result, because only 4 grams of hydrogen are available from the borohydride reagent and it appears as if 4 grams of hydrogen come from nowhere. It is here that one must note that Reaction [A3] shows that water is not a consumable reagent and that there is a net generation of water. This water can be used to supply the other half of the necessary hydrogen. Water does not need to be included in the weight of the solution, because it is intrinsic to the generator. The weight of this water would, of course, be included in the total system weight, as would be the weight of the storage tanks, reaction chambers, fittings, and the catalyst.

In current practice, the water recycle loop is not 100% efficient and the engineering difficulties make it unlikely that yields of over 20% hydrogen will be reached. The important point is that while this report describes real solutions that yield 7% hydrogen by weight, it is not inconceivable that this technology will lead to solutions that yield the energy equivalent of 10-15% hydrogen by weight. For example, 1 kilogram of slurry made with 50% sodium borohydride and 50% 1 M NaOH in aqueous solution will yield 105 grams of H_2 if enough water is recycled from Reaction [A2]. The possibility of a 10+% hydrogen solution offers an exciting future for the borohydride-water energy system.

APPENDIX B

A calculation can be done to show how much power is available from a given rate of hydrogen generation. For the anode reaction

$$H_2 \rightarrow H^+ + 2 \text{ mol e}^- \qquad [B1]$$

occurring in PEM fuel cells, the energy available from 1 mole H_2 using $\triangle G = -nFE$ is

$$(2 \text{ mol e}^-) \times (96{,}485 \text{ C /mol e}^-) \times (1.23 \text{ V}) = -2.37 \times 10^5 \text{ J / mole } H_2 \qquad [B2]$$

where the 2 moles comes from Reaction [B1], F is Faraday's constant and E comes from the combined voltage of oxidation of hydrogen [B1] and the reduction of oxygen gas (not shown). Since 1 J is equivalent to 1 W-sec,

$$(-2.37 \times 10^5 \text{ J / mole } H_2) \times (1 \text{ W-sec/J}) \times (1 \text{ hr/ } 3600 \text{ sec}) = 65.9 \text{ W-hr / mole } H_2 \quad [B3]$$

which is equivalent to

$$(65.9 \text{ W-hr / mole } H_2) \times (1\text{kW-hr / } 1000 \text{ W-hr}) = 0.0659 \text{ kW-hr / mole } H_2 \qquad [B4]$$

or 15.2 mole H_2 / kW-hr.

At standard temperature and pressure there are 24.5 liters per one mole of gas. Therefore,

$$(15.2 \text{ mole } H_2 \text{ / kW-hr}) \times (24.5 \text{ liter / mole } H_2) = 371 \text{ liters } H_2 \text{ / kW-hr} \qquad [B5]$$

This can be restated as

$$1000 \text{ W} = (371 \text{ liters } H_2 \text{ / hour}) \times (1\text{hour / } 60 \text{ minutes}) = 6.19 \text{ liters } H_2 \text{ / minute} \quad [B6]$$

In other words,

$$1 \text{ liter H2 / minute} = (1000 \text{ W}) / (12.4 \text{ liters } H_2 \text{ / minute}) = 162 \text{ Watts} \qquad [B7]$$

or 1 liter H2 / minute is sufficient hydrogen to power a 162 W fuel cell, if that fuel cell is 100% efficient.

At this point an assumption must be made as to the efficiency of the conversion of the energy available into useful energy. The efficiency of PEM fuel cells is reported to be anywhere from 30% to 60% and the mildly optimistic value of 50% has been arbitrarily selected for calculations associated with the main body of the paper. Therefore

$$1 \text{ liter H2 / minute} = (162 \text{ W}) \times (0.50 \text{ efficiency}) = 81 \text{ W fuel cell.} \qquad [B8]$$

PRODUCTION OF HYDROGEN FROM BIOMASS
BY PYROLYSIS/STEAM REFORMING

Stefan Czernik, Richard French, Calvin Feik, and Esteban Chornet

National Renewable Energy Laboratory
1617 Cole Boulevard
Golden, CO 80401

INTRODUCTION

Hydrogen is the most environmentally friendly fuel that can be efficiently used for power generation. When oxidized in a fuel cell, it produces steam as the only emission. At present, however, hydrogen is produced almost entirely from fossil fuels such as natural gas, naphtha, and inexpensive coal. In such processes, the same amount of CO_2 as that formed from combustion of those fuels is released during the hydrogen production stage. Renewable biomass is an attractive alternative to fossil feedstocks because of the potential for essentially zero net CO_2 impact. Unfortunately, hydrogen content in biomass is only 6-6.5% compared to almost 25% in natural gas. For this reason, on a cost basis, producing hydrogen by the biomass gasification/water-gas shift process cannot compete with the well-developed technology for steam reforming of natural gas. However, an integrated process, in which part of the biomass is used to produce more valuable materials or chemicals and only residual fractions are used to generate hydrogen, can be an economically viable option.

The proposed method, which was described earlier[1], combines two stages: fast pyrolysis of biomass to generate bio-oil and catalytic steam reforming of the bio-oil to hydrogen and carbon dioxide. Fast pyrolysis is a thermal decomposition process that requires a high heat transfer rate to the biomass particles and a short vapor residence time in the reaction zone. Several reactor configurations have been shown to assure this condition and to achieve yields of liquid product as high as 70-80% based on the starting biomass weight. They include bubbling fluid beds[2,3], transport reactors[4,5], and cyclonic reactors[6,7]. In the 1990s several fast pyrolysis technologies have reached near-commercial status. The concept of pyrolysis/steam reforming has several advantages over the traditional gasification/water-gas shift technology. First, bio-oil is much easier to transport than solid biomass and therefore, pyrolysis and reforming can be carried out at different locations to improve the economics. A second advantage that could significantly impact the economics of the entire process is the

Advances in Hydrogen Energy, edited by Padró and Lau
Kluwer Academic/Plenum Publishers, 2000

potential production and recovery of higher value co-products from bio-oil. In this concept, the lignin-derived fraction would be separated from bio-oil and used either directly or, after specific treatments, as a phenol substitute in phenol-formaldehyde adhesives. The carbohydrate-derived fraction would be catalytically steam reformed to produce hydrogen. Techno-economic analysis for the process was completed[8] assuming hydrogen production capacity of 35.5 t/day, which would require 930 t/day of dry biomass. Biomass was considered available at a cost of $25/dry tonne. A 15% internal rate of return was assumed for both the pyrolysis and reforming installations. If the phenolic fraction of bio-oil could be sold for $0.44/kg (approximately half of the price of phenol), the estimated cost of hydrogen from this conceptual process would be $7.7/GJ, which is at the low end of the current hydrogen selling prices.

This work focused on demonstration of the performance of a fluidized bed reactor for producing hydrogen by reforming biomass pyrolysis liquids. In previous years we showed, initially through micro-scale tests then the bench-scale fixed-bed reactor experiments,[9] that bio-oil model compounds as well as the carbohydrate-derived fraction of bio-oil can be efficiently converted to hydrogen. Using commercial nickel catalysts the hydrogen yields obtained approached or exceeded 90% of those possible for stoichiometric conversion. The carbohydrate-derived bio-oil fraction contains a substantial amount of non-volatile compounds (sugars, oligomers) that tend to decompose thermally and carbonize before contacting the steam reforming catalyst. Even with the large excess of steam used, the carbonaceous deposits on the catalyst and in the reactor freeboard limited the reforming time to 3-4 hours. For the above reasons, we decided to use a fluidized-bed reactor configuration to overcome at least some limitations of the fixed-bed unit. Though we were not able to completely prevent carbonization of the feed, the bulk of the fluidizing catalyst was in contact with the liquid droplets fed to the reactor. This greatly increased the reforming efficiency and extended the catalyst time-on-stream. Catalyst regeneration was done by steam or carbon dioxide gasification of carbonaceous residues providing additional amounts of hydrogen.

EXPERIMENTAL

The bio-oil was generated from poplar wood using the NREL fast pyrolysis vortex reactor system[10] operating at 10 kg/h of wood. The reactor wall temperature was maintained within the range of 600-625°C, which has been proven to provide the highest bio-oil yield. Nitrogen at a flow rate of 15 kg/h was used as the carrier gas for the biomass particles in the pyrolysis reactor. The yield of the bio-oil collected was 55 wt% on the dry feedstock basis. The oil was comprised of 46.8% carbon, 7.4% hydrogen, and 45.8% oxygen with water content of 19% (by weight). It was separated into aqueous (carbohydrate-derived) and organic (lignin-derived) fractions by adding water to the oil at a weight ratio of 2:1. The aqueous fraction (55% of the whole oil) contained 22.9% organics ($CH_{1.34}O_{0.81}$) and 77.1% water.

U91, a commercial nickel-based catalyst used for steam reforming of natural gas, was obtained from United Catalysts and ground to the particle size of 300-500μm.

The aqueous solution was steam reformed using a bench-scale fluidized bed reactor shown in Figure 1. The two-inch diameter inconel reactor supplied with a porous metal distribution plate was placed inside a three-zone electric furnace. The reactor contained 150-200g of commercial nickel-based catalyst of the particle size of 300-500μm. The catalyst was fluidized using superheated steam, which is also a reactant in the reforming process. Steam was generated in a boiler and superheated to 750°C before entering the reactor at a flow rate of 2-4 g/min. Liquids were fed at a rate of 4-5 g/min using a diaphragm pump. A specially designed injection nozzle supplied with a cooling jacket was used to spray liquids into the catalyst bed. The temperature in the injector was controlled by a coolant flow and maintained below the liquid boiling point to prevent evaporation of volatile and deposition of nonvolatile components. The product collection line included a cyclone that captured fine catalyst particles and, possibly, char generated in the reactor and two heat exchangers to condense excess steam. The condensate was collected in a vessel whose weight was continuously monitored. The outlet gas flow rate was measured by a mass flow meter and by a dry test meter. The gas composition was analyzed every 5 minutes by a MTI gas chromatograph. The analysis provided concentrations of hydrogen, carbon monoxide, carbon dioxide, methane, ethylene, and nitrogen in the outlet gas stream as a function of time of the test. The temperatures in the system as well as the flows were recorded and controlled by the OPTO data acquisition and control system.

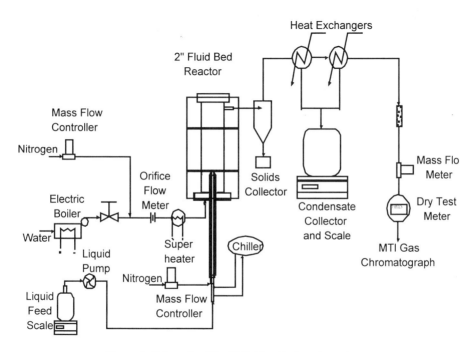

Figure 1. Schematic of the 2" fluidized bed reactor system.

The measurements allowed us to determine total and elemental balances as well as to calculate the yield of hydrogen generated from the biomass-derived liquid feed.

RESULTS AND DISCUSSION

The overall steam reforming reaction of any oxygenated organic compound can be presented as follows:

$$C_nH_mO_k + (2n-k)H_2O = nCO_2 + (2n+m/2-k)H_2$$

Thus the maximum (stoichiometric) yield of hydrogen is $2+m/2n-k/n$ moles per mole of carbon in feed. Therefore, 32.4 g of hydrogen could be theoretically obtained from 1 liter of the carbohydrate-derived fraction of the bio-oil.

The steam reforming experiments in the fluidized bed reactor were carried out at the temperature of 800°C and 850°C. The steam to carbon ratio was held at 7-9 while the methane-equivalent gas hourly space velocity $G_{C_1}HSV$ was in the range of 1200-1500 h^{-1}. At 800°C, a slow decrease in the concentration of hydrogen and carbon dioxide and an increase of carbon monoxide and methane in the product gas was observed. These changes resulted from a gradual loss of the catalyst activity, probably due to coke deposits. As a consequence of that, the yield of hydrogen produced from the bio-oil fraction decreased from the initial value of 95% of the stoichiometric potential (3.24 g of hydrogen from 100 g of feed) to 77% after 12 hours on-stream. If a water-gas shift reactor followed the reformer, the hydrogen yields could increase to 99% and 84% respectively. At a higher temperature (850°C), the product gas composition remained constant during eight hours of reforming as presented in Figure 2. This indicates that no catalyst deactivation was observed throughout the run time. The yield of hydrogen produced from the bio-oil fraction was approximately 90% of that possible for stoichiometric conversion. It would be greater than 95% if carbon monoxide underwent the complete shift reaction with steam. Only small amounts of feed were collected as char in the cyclone and condensers, and little or no coke was deposited on the catalyst.

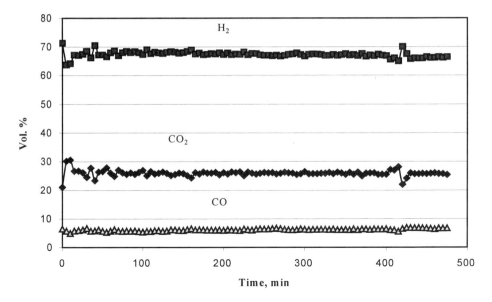

Figure 2. Reforming gas composition.

CONCLUSIONS

We successfully demonstrated that hydrogen could be efficiently produced by catalytic steam reforming of carbohydrate-derived bio-oil fractions in a fluidized bed reactor using a commercial nickel-based catalyst. Greater steam excess than that used for natural gas reforming was necessary to minimize the formation of char and coke (or to gasify these carbonaceous solids) resulting from thermal decomposition of complex carbohydrate-derived compounds.

At 850°C and a molar steam to carbon ratio of 9, the hydrogen yield was 90% of that possible for stoichiometric conversion during eight hours of the catalyst on-stream time. This yield could be 5-7% greater if a secondary water-gas shift reactor followed the reformer.

Finally, coke deposits were efficiently removed from the catalyst by steam and carbon dioxide gasification, which restored the initial catalytic activity.

ACKNOWLEDGMENTS

The authors are thankful to the U.S. Department of Energy Hydrogen Program for financial support of this work.

REFERENCES

1. Wang, D.; Czernik, S.; Montané, D.; Mann, M.; and Chornet, E., *I&EC Research*, 1997, 36, 1507.
2. Scott, D.S.; Piskorz, J.; Radlein, D., *Ind. Eng. Chem. Process Des. Dev.* 1985, 24, 581.
3. Piskorz, J.; Scott, D.S.; Radlein, D., In *Pyrolysis Oils from Biomass: Producing, Analyzing, and Upgrading;* Soltes, E.J, Milne, T.A., Eds.; ACS Symposium Series 376, ACS, Washington, D.C. 1988; pp.167-178.
4. Graham, R.G.; Freel, B.A.; Bergougnou, M.A., In *Research in Thermochemical Biomass Conversion;* Bridgwater , A.V., Kuester, J.L., Eds.; Elsevier Applied Science, London 1988; pp. 629-641.
5. Kovac, R.J.; Gorton, C.W.; O'Neil, D.J., In *Proceedings of Thermochemical Conversion Program Annual Meeting;* SERI/CP-231-2355; Solar Energy Research Institute, Golden, CO, 1988; pp. 5-20.
6. Diebold, J. and Scahill, J., In *Pyrolysis Oils from Biomass: Producing, Analyzing, and Upgrading;* Soltes, E.J, Milne, T.A., Eds.; ACS Symposium Series 376, ACS, Washington, D.C. 1988; pp. 31-40.
7. Czernik, S.; Scahill, J.; Diebold, J., *J. Sol. Energy. Eng.* 1995, 117, 2.
8. Mann, M.K.; Spath, P,L.; Kadam, K., In *Proceedings of the 1996 U.S. DOE Hydrogen Program Review*, Miami, FL, May 1-2, 1996, NREL/CP-430-21968; pp. 249-272.
9. Wang, D.; Czernik, S.; and Chornet, E., *Energy&Fuels*, 1998, 12, 19.
10. Scahill, J.W.; Diebold, J.; Feik, C.J., In *Developments in Thermochemical Biomass Conversion*, A.V. Bridgwater and D.G.B. Boocock (eds.), Blackie Academic&Professional, London 1997, pp.253-266.

EVALUATION AND MODELING OF A HIGH-TEMPERATURE, HIGH-PRESSURE, HYDROGEN SEPARATION MEMBRANE FOR ENHANCED HYDROGEN PRODUCTION FROM THE WATER-GAS SHIFT REACTION

R. M. Enick, B. D. Morreale, and J. Hill

Department of Chemical and Petroleum Engineering
University of Pittsburgh
1249 Benedum Hall, Pittsburgh, PA, 15261

K. S. Rothenberger, A. V. Cugini, R. V. Siriwardane, and J. A. Poston

U.S. Department of Energy
National Energy Technology Laboratory
P.O. Box 10940, Pittsburgh, PA 15236-0940

U. Balachandran, T. H. Lee,and S. E. Dorris

Energy Technology Division
Argonne National Laboratory
Argonne, IL 60439

W. J. Graham and B. H. Howard

Parsons Project Services Incorporated
National Energy Technology Laboratory
Library, PA 15129

ABSTRACT

A novel configuration for hydrogen production from the water gas shift reaction is proposed, using high temperature to enhance the rate of reaction and employing a hydrogen separation membrane for the collection of the high purity hydrogen product. Experimental high-temperature, high-pressure flux measurements have been made on a material that shows promise for such an application. Mixed oxide and metal "cermet" ion-transport disk membranes, fabricated at Argonne National Laboratory, were evaluated for hydrogen permeability on a unique high-temperature, high-pressure test unit constructed at the

National Energy Technology Laboratory. Hydrogen permeation was found to be proportional to $\Delta P_{H2}^{0.5}$. At 700°C, the membrane permeability was 9.62×10^{-3} [cm^2/min][mol/lit]$^{0.5}$, or 5.63×10^{-9} [mol/m s Pa$^{0.5}$]. The membrane permeability increased to 3.31×10^{-2} [cm^2/min][mol/lit]$^{0.5}$, or 1.76×10^{-8} [mol/m s Pa$^{0.5}$] at 900°C. The membrane material was also characterized for surface changes and structural integrity using scanning electron microscopy/X-ray microanalysis, and X-ray photoelectron spectroscopy as a function of temperature, pressure, and hydrogen exposure. Although the membrane performed well for the short periods of time employed in this study, long-term stability remains a concern. The feasibility of using this mixed-oxide ceramic membrane to remove hydrogen from the reaction mixture was modeled for the optimization of the water-gas shift reaction. This reversible reaction is characterized by a very low equilibrium constant at elevated temperatures (>800°C); consequently CO conversion at these temperatures is typically less than 50%. In the scenario modeled, CO conversion was increased from 35% in the absence of a membrane to 79% with a membrane present, with still higher values possible if hydrogen was actively removed from the permeate side of the membrane. In the model, the effectiveness of the configuration is limited by the buildup of hydrogen partial pressure on the permeate side of the membrane. The model provided estimates of the conversion of CO attained for a specified feed, reactor size and permeate pressure.

INTRODUCTION

The demand for hydrogen is expected to rise in coming years with increases in its use both directly as a fuel and indirectly in the synthesis or upgrading of fuels required to meet increasingly demanding environmental standards. Industrial processes used to produce hydrogen, such as steam reforming of natural gas, or the gasification of heavy carbonaceous materials, generally proceed at elevated temperatures and/or pressures and produce hydrogen mixed with other gases. The recovery of hydrogen from high-temperature, high-pressure streams without the need for cooling or depressurization would enhance the efficiency and performance of systems based on such streams. Advances in the area of membrane technology may provide the basis for improved methods of hydrogen recovery and thus reduce the cost associated with hydrogen production at elevated temperature and pressure.

The water-gas shift reaction is important as a method for further enhancing the yield of hydrogen from such aforementioned industrial processes. However, the production of hydrogen via the water-gas shift reaction is favored at low temperatures. Therefore, the common approach when using the water-gas shift reaction has been to decrease the temperature of the process to favor the formation of hydrogen, and to employ a catalyst to enhance the kinetics of reaction. If the reactant gases were produced at high temperature, a loss of efficiency results upon lowering the temperature. In addition, the need for a catalyst increases costs and introduces an added degree of complexity into the reaction process.

In this paper, we will present a model of a novel configuration for hydrogen production from the water-gas shift reaction. This configuration will involve running the water-gas shift at high temperature while employing a hydrogen separation membrane for collection of the hydrogen product [Enick et al., 1999]. The high temperature will improve overall process efficiency while obviating the need for a catalyst to speed the reaction. The membrane will shift the otherwise unfavorable equilibrium towards the production of hydrogen. In addition, we will add credence to the feasibility of such a configuration by presenting experimental hydrogen flux data taken at high temperatures and pressures on small disk samples of a proton conducting ceramic-metal material. Such a material, though still in the early stages of development, has the potential to serve as a hydrogen separation membrane in high severity process environments.

The Water-Gas Shift Reaction

The water-gas-shift reaction (Eqn. 1) has been studied extensively as a basis for improving the yield of hydrogen production. In many applications, including ammonia synthesis or fuel reforming for proton exchange membrane (PEM) fuel cells, the maximum acceptable level of CO in hydrogen is in the parts per million range, therefore the water-gas shift reaction is needed.

$$CO + H_2O \Leftrightarrow CO_2 + H_2 \tag{1}$$

There is no change in the number of moles as the reaction proceeds; therefore, the equilibrium conversion is not affected by pressure. Side reactions associated with the water-gas-shift reaction are usually not significant.

The equilibrium constant for this exothermic reaction (K, which can be expressed in terms of the concentrations of the reactants and products, Eqn. 2) decreases as temperature increases. For example, the value of K decreases from 3587 at 373 K to 0.5923 at 1273 K, as found in a published table of equilibrium values [Rostrup-Nielsen, 1984].

$$K = \frac{[CO_2][H_2]}{[CO][H_2O]} \tag{2}$$

Therefore, conversion of CO to CO_2 (low concentrations of CO and H_2O, high concentrations of CO_2 and H_2) is favored at low temperature [Benson, 1981]. The low equilibrium conversion of CO to CO_2 at elevated temperatures is the major reason that most theoretical, experimental and industrial studies of the water-gas shift reaction are conducted at low to moderate temperatures.

Kinetics of the Water-Gas Shift Reaction

Most heterogeneous catalysis studies of the water-gas shift reaction have been conducted at temperatures less than 450°C. Examples of commercial water-gas shift catalysts include Fe_3O_4-Cr_2O_3 and $CuZnO/Al_2O_3$ [Keiski et al., 1993]. The kinetics associated with these catalysts can be adequately described with pseudo-first order or power law kinetics [Keiski et al., 1993]. Another study of the stationary and transient kinetics of this reaction [Keiski et al., 1996] indicated that various mechanisms and kinetic expressions have been proposed for the water-gas shift reaction, and that Langmuir-Hinshelwood and power-law kinetic models are adequate. An earlier study of the water-gas shift reaction over a chromia-promoted iron oxide catalyst not only accounted for temperature and pressure, but also the catalyst age, diffusion and gas-phase composition including H_2S concentration [Singh and Saraf, 1977]. The water-gas shift reaction also occurs in some processes that do not employ catalysts. For example, the supercritical water oxidation of organic wastes (typically conducted at 400-550°C, 200-300 bar) usually does not employ a catalyst because of the rapid destruction rates that are achieved. The very high rate of the water-gas shift reaction observed in this system [Helling and Tester, 1987; Holgate et al., 1992; Rice et al., 1998] was attributed to the formation of "cages" of water about the reactants under supercritical conditions and very high water concentrations.

There is relatively little information on the kinetics of the water-gas shift reaction at elevated temperatures (>600°C). This can be primarily attributed to the diminished value of K, which would limit CO conversion to unacceptably low levels in conventional reactors. Catalysts are typically not used at elevated temperatures because of the rapid rate of the non-catalyzed reaction and catalyst instability at these extreme conditions. A study of the forward and reverse reactions of the water-gas shift reaction (Eqn. 1) was conducted at extremely high temperatures and low pressure (800-1100°C, 1 bar) [Graven and Long, 1954]. No catalyst was employed and power-law kinetic expressions were developed for the forward and reverse reactions. Despite the rapid attainment of equilibrium conversions without a catalyst, the ability to convert high-temperature (800-1000°C), CO-rich combustion gases into hydrogen fuel via the water-gas shift reaction is limited by the low equilibrium conversions of CO.

Enhancing the Production of Hydrogen by Hydrogen Removal with Membranes

It has long been recognized that high levels of conversion in equilibrium-limited reactions can be achieved if one or more of the products can be extracted during the reaction. For a given value of the equilibrium constant in Eqn. 2, if the concentration of either CO_2 or H_2 is reduced the concentrations of CO and H_2O must also decrease, thereby increasing the conversion of CO. Previous investigators have established that hydrogen-permeable membranes provide a means of increasing hydrogen production in the water-gas shift reaction, steam reforming reactions, or other reversible reactions that produce hydrogen, such as the dehydrogenation of ethylbenzene or cyclohexane. The use of catalytic or inert hydrogen permeable membranes in a reactor would result in the removal of hydrogen from the reaction mixture, leading to an increase in feedstock conversion in the equilibrium-limited reactions yielding hydrogen as a product [Armor, 1995; Sun and Khang, 1988; Liu et al., 1990; Barbieri and DiMaio 1997a,b; Uemiya et al., 1991a,b; Tsotsis et al., 1992; Itoh et al., 1993; Adris et al., 1994; Netherlands Energy Research Foundation, 1995; Basile et al., 1996a,b; Mardilovich et al., 1998]. In this study, only the water-gas shift reaction was modeled.

Membrane Characterization

Currently, several research organizations are engaged in the development of hydrogen transport membranes or their precursor materials [Balachandran et al., 1998; Fain and Roettger 1993; Peachey et al., 1999; Mardilovich et al., 1998]. Membrane materials range from organic polymers to metals to ceramics. Non-porous (high density) ceramic membranes are particularly desirable because they can be made exclusively selective to hydrogen and are durable enough to withstand high temperature, high pressure conditions that would be encountered in the proposed commercial application [Balachandran et al., 1998]. Practical application of these membranes would likely employ a high inlet pressure coupled with reduced pressure on the outlet side of the membrane to enhance the flux. Hydrogen flux through these membranes is expected to be optimal in the range of 700-900°C and increase with an increasing hydrogen partial pressure gradient across the membrane. The proton-conducting ceramic membranes that were used in this study were developed at Argonne National Laboratory (ANL). These dense ceramic membranes were fabricated from mixed protonic/electronic conductors and had previously been tested at ANL under ambient pressures. Hydrogen selectivity was very high because the membranes do not have interconnected pores; the only species that pass through the membranes are those that participate in ionic conduction, such as hydrogen [Balachandran et al., 1998]. These membranes had not been previously tested at elevated pressures or over a wide range of hydrogen concentrations.

The goal of this study was the characterization of the ANL ion-transport membrane at elevated temperatures (700-900°C) and pressures ($\Delta P_t < 250$ psi) that were relevant to the water-gas shift reaction concept. Hydrogen permeability was evaluated using steady state measurements of hydrogen flux through the membrane. Characterization of surface changes and structural integrity were performed using scanning electron microscopy/X-ray microanalysis (SEM/EDS), X-ray photoelectron spectroscopy (XPS) and atomic force microscopy (AFM), as a function of temperature and hydrogen exposure. Steady-state hydrogen flux measurements over the 700-900°C range at hydrogen partial pressure differences up to 250 psi were used to characterize hydrogen permeability. The expression derived for the hydrogen permeability of the membrane was then included in the water-gas shift membrane-reactor model.

EXPERIMENTAL METHODS

Cermet membranes, consisting of a metal oxide base mixed with a metal, were fabricated by a process developed at ANL. The membranes used in this study were of composition "ANL-1", with an oxide of composition $BaCe_{0.80}Y_{0.20}O_3$ (BCY) with nickel as the added metal. The BCY ceramic base was prepared by mixing appropriate amounts of $BaCO_3$, CeO_2 and Y_2O_3, then calcining the mixture at 1000°C for 12 hours in air. The powder was then ball-milled and calcined again at 1200°C for 10 hours in air. After obtaining phase-pure powder (as determined by X-ray diffraction), the BCY powder was mixed with 40 vol% metallic nickel powder to increase its electronic conductivity. The powder mixture was then uniaxially pressed and sintered for 5 hours at 1400-1450°C in an atmosphere of 4% hydrogen/96% argon. The disk membranes were typically 0.04 - 0.06 in (1.0 - 1.5 mm) in thickness.

Membranes for pressure and flux testing were mounted, using a brazing process developed at ANL, within four inch long, 0.75 in. O.D., heavy-wall, Iconel 600 tubing which had been machined to form a small seat to accommodate the membrane. Typically, the membrane diameter was 0.69 in. (17.5 mm.). Unmounted membranes of the same composition, as well as membranes of varying Ni content, were also available for characterization studies. Because the pressure tested membranes had to be pre-mounted, the before-and-after characterization studies refer to membranes of the same composition and fabrication, but not the same physical membranes.

Membrane pressure and flux testing was performed using the Hydrogen Membrane Testing (HMT) unit which was constructed in the Hydrogen Technology Research (HTR) facility at the National Energy Technology Laboratory (NETL). The facility made use of the high pressure hydrogen handling infrastructure previously put in place for the study of high pressure hydrogenation reactions [Cugini et al., 1994]. For membrane testing, the unit has an operating range from ambient temperature to 900°C and ambient pressure to 450 psig. The Inconel 600 tubing containing the pre-mounted membrane from ANL was welded to an additional length of 0.75 in. O.D. Inconel 600 tubing. The membrane was hung in an inverted configuration and attached to a second piece of 0.75 in. O.D. Inconel 600 tubing by means of an Inconel 600 sleeve. A ceramic fiber heater was positioned around the sleeve for heating the membrane. The entire assembly was suspended within a 2-gallon stainless steel autoclave that was continuously purged with nitrogen gas. A simplified schematic of the test assembly is shown in Figure 1.

Both cold and hot inert gas pressure tests were performed on samples of brazed cermet membranes in advance of hydrogen flux measurements in order to ascertain the integrity of the Inconel 600-to-cermet seal under high temperature and pressure conditions. The procedure used in these tests has been described elsewhere [Rothenberger et al., 1999].

Hydrogen flux measurements were performed on three separate ANL-1 cermet membranes. Three different test gases - 100% hydrogen, a 50% hydrogen/50% nitrogen mixture, and a 10% hydrogen/90% nitrogen mixture - were introduced in a flowing mode to the retentate (feed) side of the membrane. The flow rate, typically 190 ml/min STP, was sufficient such that the overall composition of the retentate side did not change significantly due to hydrogen permeation. Operating temperatures ranged from 700-900°C as measured by a thermocouple positioned on the permeate side of the membrane, approximately 0.25 in. above the membrane surface (Figure 1). Test gas pressures were stepped both up and down in 10-50 psi increments between atmospheric pressure and 250 psig. Pressure conditions were held from one to four hours, depending on the magnitude of the step from the previous reading, and whether the flux data had reached a constant value. Usually, the flux data reached a constant value within 10-20 minutes. The permeate-side of the membrane was swept with argon, and the effluent was monitored with a gas chromatograph (GC) for hydrogen concentration using an argon carrier gas. GC measurements were recorded every 10 minutes during flux testing. The argon sweep rate was maintained at 85 ml/min STP, resulting in hydrogen concentrations in the sweep gas in the range of 0.01 to 3.0 percent. Nitrogen detected by GC was indicative of a membrane leak. Although the experiments were allowed to continue with a leaking membrane, only data taken before the detection of a leak was used in the membrane permeance evaluation for this paper.

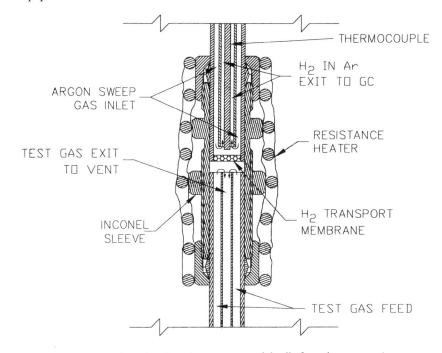

Figure 1. Schematic of membrane reactor and detail of membrane mounting.

Because the Inconel 600 metal alloy used as a mounting material for the membrane itself becomes permeable to hydrogen under the test conditions, a separate experiment, configured in a similar manner, was performed to determine the permeance of the Inconel 600.

X-ray photoelectron spectroscopy (XPS) spectra were recorded with a Physical Electronics model 548 XPS system. The binding energies were referenced to the C(1s) level at 284.6 eV for adventitious carbon. XPS data were obtained at various temperatures ranging from room temperature to 650°C. X-ray microanalysis was performed at room temperature and 575°C using a JOEL 840-A scanning electron microscope equipped with a Noran Instruments Micro-Z energy dispersive spectrometer, which was interfaced to a Noran Instruments Voyager-4 computer. Detector resolution, as referenced to the manganese Kα spectra line, was 148 eV.

Data obtained at NETL for hydrogen flux through the ANL-1 dense ceramic membranes were correlated to yield a flux expression. The flux of hydrogen through these membranes, R_{H2}, was found to be proportional to the permeability of the membrane, k_{H2}, and was assumed to be inversely proportional to the membrane thickness, t_m, as is typical with other hydrogen permeable membrane materials. The driving force for the hydrogen flux was related to the hydrogen partial pressure, P_{H2}, or molar concentration of hydrogen, C_{H2}, raised to the exponent n (the units of k are dependent upon the value of n), with the units of R'_{H2} (mol/s), R_{H2} (mol/s), A_m (m^2), t_m (m), C_{H2} (mol/L), and P_{H2} (Pa).

$$R'_{H2} = \frac{A_m k'_{H2} \left(C_{H2,retentate}{}^n - C_{H2,permeate}{}^n \right)}{t_m} * 1.67 \times 10^{-3} \tag{3}$$

$$R_{H2} = \frac{A_m k_{H2} \left(P_{H2,retentate}{}^n - P_{H2,permeate}{}^n \right)}{t_m} \tag{4}$$

Now, flux will be equivalent regardless of the form of the driving force, therefore

$$R'_{H2} = R_{H2} \tag{5}$$

If transport through a membrane involving both surface reaction (dissociation) and diffusion was limited by surface reactions, then n = 1. If transport was diffusion-limited, then n = 0.5. Intermediate values of n (0.5<n<1) have also been reported [Uemiya et al., 1991a; deRossett 1960; Hurlbert and Konecny, 1961]. In this study, the flux of hydrogen through the proton transport membranes was modeled with the same form of the equations used to model diffusion membranes, Eqns. 3 and 4. Values of k'_{H2} and n were determined from Eqn. 3. The concentration of hydrogen on the permeate-side was insignificant relative to the concentration on the retentate-side. Therefore $C_{H2,permeate}$ was equated to zero. Thus, taking the logarithm of both sides of Eqn. 3 yields,

$$\log R_{H2} = n \log C_{H2,retentate} + \log(A_m k'_{H2} / t_m) \tag{6}$$

Therefore, a log-log plot of hydrogen permeation through the membrane, R_{H2} [mol/min] versus retentate hydrogen concentration, $C_{H2,permeate}$ [mol/cm^3], yielded a straight line of slope n. The value of k'_{H2} was determined from the intercept because both membrane area, A_m, and membrane thickness, t_m, were known. Membrane areas were in the 1.25-1.36 cm^2 range, depending on the specific to wall thickness used for membrane mounting, and membrane thickness was in the 1.1-1.3 mm range.

Subsequent studies of these proton-transport membranes will be conducted over a wide range of hydrogen partial pressures on the permeate side. This will allow assessment of whether the driving force for hydrogen permeance is more accurately represented as the natural logarithm of the ratio of the hydrogen partial pressures on the retentate and permeate sides, as observed in ion transport oxygen membranes.

RESULTS AND DISCUSSION

Membrane Characterization

The cold and hot inert gas pressure tests on the sealed ANL-1 membranes are described elsewhere [Rothenberger et al., 1999]. The results indicated that the membrane to substrate seal was gas tight to 400 psig at temperatures up to 800°C, at least for short periods of time. However, a small leak developed at a temperature of 800°C and a pressure of 450 psig. After cool down, visual inspection of the membrane revealed a white powdery coating on and around the membrane. Small areas of green discoloration were observed around the surface of the normally gray membrane. Some of the brazing material appeared to have migrated from around the edge of the membrane toward the center, moving a distance of approximately 0.5 mm. Migration of the braze material had previously been observed during hot ambient temperature flux testing at ANL. Visual inspection under an optical microscope revealed cracks in the membrane surface along the perimeter of the disk, as well as cracks in the brazing material itself. At this time, it is not known whether the failure was strictly a pressure effect, or if the elevated temperature and possible migration of some of the brazing material was involved. Because the hydrogen membrane testing unit was still under construction at the time of the pressure tests, the permeate side of the membrane was exposed to atmospheric oxygen during the inert gas pressure tests. The membrane material is also permeable to oxygen at high temperature, and the presence of what is probably oxide contamination around the membrane after the hot pressure test indicated that some degradation of the membrane and/or sealing materials may have occurred as a result of air exposure. Following the inert gas pressure tests, it was decided to limit the pressure drop across the membrane during flux testing to 250 psi.

Hydrogen flux testing of the ANL-1 cermet membranes was conducted on the HMT unit. Data from these runs is plotted in Figure 2.

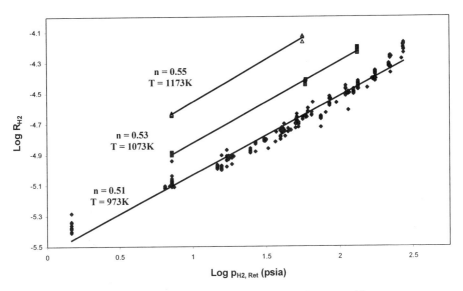

Figure 2. Hydrogen permeation data as a function of hydrogen partial pressure.

The unit behaved as expected, with increased hydrogen flux observed as a function of increased hydrogen partial pressure and increased membrane temperature from 700°C to 900°C. Operational difficulties were encountered in maintaining feed gas flow rate at pressures approaching 250 psig, resulting in drifting or poorer quality data at those values. A small systematic bias was also observed resulting in slightly (approximately 5%) higher flux readings detected when stepping down to a given pressure than when stepping up to the same pressure. However, no attempt was made to select or discard individual data points. All data points taken during a given test period were used in the assessment of membrane performance. For the three membranes that were tested, leakage was detected after approximately 70, 50, and 20 hours of use, respectively. No data taken after the detection of any leak was used in the permeance analysis. Physical and microscopic examination of the membranes after testing revealed that the leak was generally located in the area of the ceramic to metal seal, rather than in the body of the membrane itself.

Hydrogen flux measurements made on the Inconel 600 mounting material revealed permeance values between 1 and 10 percent of those measured for the cermet membrane mounted in the Inconel 600. The measured hydrogen flux for the mounted membrane includes a flux contribution from the Inconel 600 holder. Unfortunately, it was impossible to make a direct measurement of hydrogen bypassing the membrane during a flux test. But in order to bypass the membrane, hydrogen permeating through the Inconel 600 mount had to diffuse first into the very tight space between the Inconel 600 sleeve and the mounting tube. Although hydrogen gas is certainly capable of this, the amount of hydrogen diffusing into the sweep zone via this route would have been limited by mass transport considerations, and thus, smaller than the amount experimentally measured. Therefore, it was decided not to make any adjustments or corrections to the membrane permeance data at this time.

SEM microanalysis was conducted to determine the changes in morphology, elemental distribution and composition that occur to a fresh ANL-1 membrane upon heating. A photomicrograph of a fresh membrane is shown in Figure 3a. No major morphological changes were observed after heating the membrane from room temperature to 575°C under an inert atmosphere. Elemental distribution was uniform and remained uniform following heating and hydrogen exposures at 575°C. SEM analysis of a 40% Ni /BCY membrane after flux testing revealed quite different results. Morphology more closely resembled the membrane samples with much lower Ni contents. X-ray microanalysis confirmed a lower Ni content in the near surface region as compared to the fresh membrane. A photomicrograph of a membrane after flux testing is shown in Figure 3b.

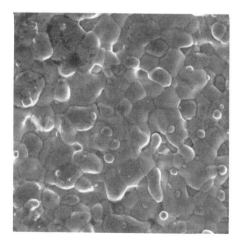 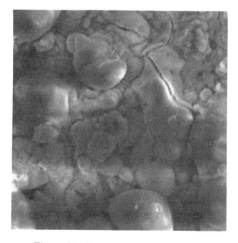

Figure 3a. Pre-run membrane SEM photomicrograph (Magnification 2K).

Figure 3b. Post-run membrane SEM photomicrograph (Magnification 2K).

XPS was utilized to determine the elemental composition and oxidation states of elements in approximately the top 50 Å of the surface of a fresh membrane during heating. At room temperature, surface nickel was oxidized and the intensity of the nickel peak was low. When the membrane was heated to 650°C under vacuum, the intensity of the nickel peak increased substantially and the nickel was reduced to the metallic state. The ratios of Ni/Ba, Ni/Ce and Ni/Y at the surface increased when the temperature was increased, as shown in Table 1, but they decreased again when the membrane was cooled back to room temperature. Thus, the nickel "migrated" or otherwise became more apparent at the surface and preferentially resided there relative to the other elements at higher temperature.

Table 1. Relative elemental composition at the surface of the fresh membrane and after H_2 flux testing

Condition	Ni / Ba	Y / Ba	Ce / Ba
Fresh at 25 °C	0.17	0.75	0.28
Fresh at 650°C	0.80	0.38	0.22
After H_2 test at 25°C	0	0	0
After H_2 test at 650°C	0.02	0.04	0

XPS analysis was also performed on the membrane following the flux test. At room temperature, only Ba was detected on the surface of the tested membrane. Ni, Ce, and Y could not be detected (see Table 1). When the sample was heated to 650°C under vacuum, it was possible to detect a small quantity of Ni and Y, but no Ce. Ba had the highest concentration at the surface both at room temperature and 650°C. The relative concentration of Ni at the surface was significantly lower on the tested membrane than on the fresh membrane. The small amount of Ni observed at 650°C on the tested membrane did not appear to be in the metallic state. The XPS results indicated that there was a permanent loss of Ni from the surface region over the course of flux testing.

Although the membrane and sealing methodology are impermeable to nitrogen and helium and are capable of withstanding high temperatures and high-pressure differentials for short periods of time, the seal appears to degrade over a 50-100 h time period at elevated temperature under any pressure. In addition, surface characterization results indicate that irreversible changes occur to the membrane itself. These results call into question the membrane seal applicability for long-term use in harsh operating environments and provide a challenge for future improvements in the technology.

Hydrogen Permeance

Values of k_{H2} and n at elevated temperatures and pressures were determined from hydrogen flux data. The log-log representation of this data (Eqn. 6) in terms of hydrogen partial pressure is illustrated in Figure 2. The n values were 0.51, 0.53 and 0.55 at 700, 800 and 900°C, respectively. This value of approximately 0.5 is similar to the values reported for diffusion membranes such as palladium [Armor, 1995]. The temperature-dependent k_{H2} value was determined from the intercept of each set of data. The results were then fit to an Arrhenius relation (Eqn. 7),

$$k'_{H2} = k_o \exp\left[\frac{-E}{8.314T}\right] \tag{7}$$

with a membrane activation energy value, E, of 58,600 J/mol, temperature in Kelvin and a pre-exponential constant value, k_o, of 13.4 $[cm^2/min][mol/liter]^{0.5}$. For example, at 700°C (973 K), the permeability of the membrane, k'_{H2}, was 9.62×10^{-3} $[cm^2/min][mol/liter]^{0.5}$, which corresponded to a k_{H2} value of 5.63×10^{-9} $[mol/(m\ s\ Pa^{0.5})]$. The membrane permeability increased to 3.31×10^{-2} $[cm^2/min][mol/lit]^{0.5}$, or 1.76×10^{-8} $[mol/(m\ s\ Pa^{0.5})]$ at 900°C. To facilitate comparison of hydrogen permeability of the ANL-1 proton conducting membrane and metal membrane permeability [Buxbaum and Marker, 1993], it is convenient to express the permeability with units of k_{H2} $[mol/(m\ s\ Pa^{0.5})]$. Conversion between k'_{H2} and k_{H2} is provided below (Eqn. 8), with temperature in Kelvin.

$$k_{H2} \frac{mol}{msPa^{0.5}} = 1.828 \times 10^{-5} \frac{1}{T^{0.5}} k'_{H2} \frac{cm^2}{min} \sqrt{mol\big/l} \tag{8}$$

At 700°C, the permeability of palladium is approximately 3.0×10^{-8} $[mol/(m\ s\ Pa^{0.5})]$ and the permeability of iron is 5.0×10^{-10} $[mol/(m\ s\ Pa^{0.5})]$ [Buxbaum and Marker, 1993]. Figure 4 illustrates the hydrogen permeability of palladium, iron and the proton conducting ceramic material as a function of inverse temperature. The driving force for hydrogen flux in each membrane material is $\Delta P^{0.5}$.

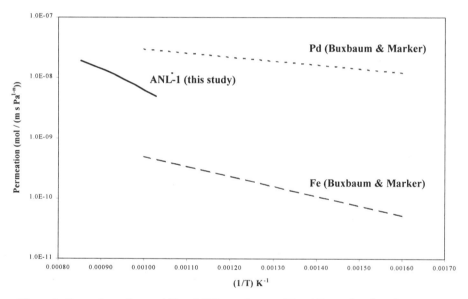

Figure 4. Comparison of permeability of ANL membrane to Pd and Fe as a function of temperature.

Modeling

Membrane reactor models of various configurations, complexity, and ranges of applicability have been previously reported [Sun and Khang, 1988; Itoh and Govind, 1989; Liu et al., 1990]. Several previous investigators have presented water-gas shift membrane reactor models. A model of the iron-chromium oxide catalyzed water-gas shift reaction at 673 K in a cylindrical, palladium membrane reactor was developed to demonstrate enhanced CO conversion [Uemiya et al., 1991a]. A packed bed catalytic membrane reactor model was used to get good agreement with experimental data by varying a single parameter, the thickness of the permselective membrane layer [Tsotsis et al., 1992]. A simple model for the water-gas shift membrane reactor was developed that assumed the reaction kinetics were so fast that equilibrium was established along the entire reactor length [Damle et al., 1994]. An ideal flow model was derived for an isothermal, cylindrical, catalyzed, water-gas shift, membrane reactor operating at temperatures up to 400°C [Netherlands Energy Research Foundation, 1995]. The same authors also used computational flow dynamics software to model the same system under either isothermal or adiabatic conditions which accounted for two-dimensional flow and axial and radial dispersion.

Unlike these previous models, the objective of this study is to model a non-catalyzed, high temperature, proton-conducting membrane reactor that incorporates reaction kinetics. Therefore, the high temperature, high pressure hydrogen permeance of the ANL-1 proton-conducting membranes was integrated into a water-gas shift membrane reactor model in this study, along with previously published high temperature, non-catalyzed reaction rate results [Graven and Long, 1954]. The configuration of the reactor was similar to the multiple, parallel tubular membranes envisioned for commercial applications [Parsons, 1998]. The model was used to estimate the permeate pressure and/or surface area of the membrane required to achieve a desired level of CO conversion or hydrogen recovery. Alternately, it was used to provide estimates of the hydrogen permeability the membrane must exhibit to attain a specified CO conversion or H_2 recovery in a reactor with a specified membrane area.

The basis of the model is a tubular membrane located within a coaxial cylindrical shell. The feed gases are introduced on the shell side (reaction side, annular side, retentate side, raffinate side) of the reactor. Hydrogen permeates the membrane, and the hydrogen permeate product can be obtained from the tube side. A schematic of the modeled membrane reactor is shown in Figure 5. As the reaction gases proceed down the length of the reactor, hydrogen will continue to permeate the membrane if the partial pressure of hydrogen on the reaction side exceeds the permeate hydrogen partial pressure, increasing the conversion of the CO. The high-pressure CO_2-rich retentate exits the reactor on the shell side, and a low-pressure, high-purity hydrogen permeate is recovered from the tube side. Use of a sweep gas is not desirable to avoid subsequent separations of hydrogen from sweep gas. Absence of a sweep gas will result in a substantial pressure drop across the wall of the membrane.

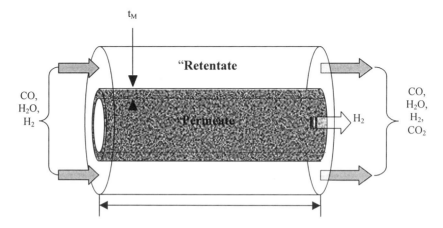

Figure 5. Schematic of reactor as modeled.

The design equations for a tubular plug flow reactor are based on ordinary differential mass balances for each component [Fogler, 1992; Netherlands Energy Research Foundation, 1995]. The model predicts the performance of the reactor under steady-state isothermal conditions. The reactants and products are assumed to behave as ideal gases. The pressure drop along the length of the reactor is assumed to be negligible for both the retentate side and the permeate side of the reactor. Published correlations for the equilibrium constant (K) [Benson, 1981; Netherlands Energy Research Foundation, 1995; Singh and Saraf, 1977] can be used for pseudo-first order or power-law kinetic expressions. Reaction kinetics are described by non-catalytic, high temperature 800-1100°C expressions [Graven and Long, 1954]. The results of their model are summarized by reaction rate expressions for the forward and reverse reactions.

$$-\frac{d[H_2O]}{dt}\quad\frac{mol}{min\;liter}=\frac{60(9.5\times10^{10}\exp(-57000/1.987T)[H_2]^{0.5}[CO_2])}{(1+3.6\times10^3[CO])} \tag{9}$$

$$-\frac{d[CO_2]}{dt}\quad\frac{mol}{min\;liter}=\frac{60(5.0\times10^{12}\exp(-67300/1.987T)[CO]^{0.5}[H_2O])}{(1+1.2\times10^4[H_2])^{0.5}} \tag{10}$$

As Graven and Long noted, these expressions were based on conditions removed from equilibrium. If the forward and reverse reaction rates are equated to determine the equilibrium constant, there will be disagreement between the result and well-established correlations for the equilibrium constant. Therefore, we modified their model in the following manner to yield the correct equilibrium constant.

$$-\frac{d[CO_2]}{dt}\quad\frac{mol}{min\;liter}=\frac{60(5.0\times10^{12}\exp(-67300/1.987T)[CO]^{0.5}[H_2O])}{(1+1.2\times10^4[H_2])^{0.5}}(1-\beta) \tag{11}$$

$$\beta=\frac{[CO_2][H_2]}{[CO][H_2O][K_{eq}]} \tag{12}$$

A recent correlation for the equilibrium constant was used [Netherlands Energy Research Foundation, 1995].

$$K_{eq}=\exp\left\{-4.33+\frac{4577.8}{T(K)}\right\} \tag{13}$$

The conditions listed in Table 2 are associated with a conceptual coal processing plant for producing hydrogen while recovering carbon dioxide [Parsons 1999]. The highest possible conversion of CO for the water-gas shift reaction in the absence of a membrane is 35%. The single tube reactor, Figure 5, can represent many parallel reactor membrane tubes within a large reactor shell.

The membrane reactor concentration profiles and molar flow rate profiles are illustrated in Figures 6 and 7, respectively. As the CO and H₂O enter, the CO concentration in the annulus is greater than the equilibrium concentration and the reaction proceeds rapidly. Hydrogen permeation also initiates. The reaction side molar flow rate of hydrogen increases initially because the rate of hydrogen generation exceeds the permeation rate. The equilibrium conversion is attained 50 cm down the length of the reactor, but the conversion of CO continues as hydrogen permeates the membrane. In the middle portions of the reactor, more hydrogen is permeating through the membrane than is being generated by the water-gas shift reaction, as a result the hydrogen concentration of the retentate decreases. Near the end of the reactor, the hydrogen concentration in the permeate and retentate approach the same value, and further conversion of CO cannot be achieved because hydrogen permeation ceases and the CO concentration is at its equilibrium value. This reactor achieved a 78.7% conversion of CO.

Table 2. Model conditions and assumptions used in the single tube example

Catalyst	No catalyst used at this high temperature
Mode of Operation	Isothermal
Temperature	900°C; 1173 K
Reaction side	Annular side, Shell side
Pressure on reaction side	27 atm
Pressure on tube side	1.0 atm
Reactor length	700 cm
Reactor diameter (for each tube)	5.0 cm
Tube-side sweep gas	None
Membrane	Proton Transport
Membrane diameter	2.0 cm
Permeable gases	Hydrogen Only
Membrane thickness	0.1 cm (1 mm)
Membrane driving force	$C_{H2,retentate}^{0.5} - C_{H2,permeate}^{0.5}$
Membrane permeability	0.033 [cm²/min][mol/liter]$^{0.5}$ (from Eq. 7)
CO inlet flow rate	0.080 gmol/min
Steam inlet flow rate	0.088 gmol/min
Hydrogen inlet flow rate	0.040 gmol/min
Reaction kinetics	Modified Graven and Long Model

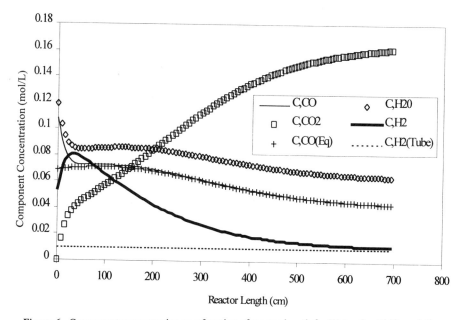

Figure 6. Component concentration as a function of reactor length for Water-Gas Shift modeling.

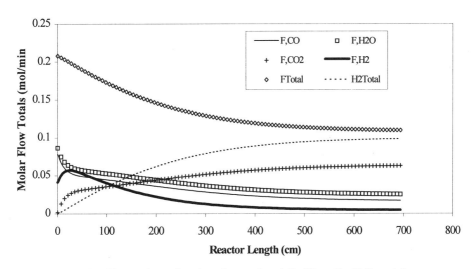

Figure 7. Flow totals as a function of reactor length for Water-Gas Shift modeling.

Significant increases in equilibrium conversion beyond 78.7% can be attained in this example only if the partial pressure of hydrogen in the permeate region is reduced. This can be accomplished by operating the permeate side of the reactor under partial vacuum or by employing a sweep gas that can be readily separated from the hydrogen in a subsequent processing step. For example, if superheated steam is used as a sweep gas, the permeate product could be cooled, condensing the water and yielding high purity hydrogen. Figure 8 illustrates the effect of hydrogen permeate partial pressure on the conversion of CO in this reactor.

CONCLUSIONS

A model of a novel configuration for hydrogen production from the water-gas shift reaction has been demonstrated. The model employs a hydrogen separation membrane for collection of the hydrogen product and high temperature for enhancing the rate of reaction. Experimental high-temperature, high-pressure flux measurements have been made on a novel material that might be suitable in such an application.

The ANL-1 cermet material tested exhibits many positive qualities for use as a hydrogen separation membrane in high-temperature, high-pressure applications. It exhibits very high selectivity, reasonable permeability, and is quite robust, being able to tolerate temperatures up to 900°C and pressure drops up to 400 psi. However, membrane-to-substrate seal methodology is an area for improvement, as seal integrity for membranes tested both at NETL and ANL was limited to durations of less than 100 hours. Both visual and spectroscopic investigation of membranes after testing suggested that the sealing and membrane materials may have reacted. A possible way to mitigate this situation might be to remove the seal from the hot zone.

Membrane surface characterization and analysis indicated that significant elemental and oxidative changes occurred on the membrane surface upon heating, and again after flux testing. In particular, XPS indicated that metallic Ni "migrated" or otherwise appeared at the surface of a fresh membrane at elevated temperature. Some of this nickel could have migrated from the braze material. After testing, both XPS and SEM/EDS indicated a loss of Ni in the surface and near surface regions. Therefore, transport of hydrogen through the membrane over an approximately 50-hour period appeared to result in permanent changes to the membrane. The effect of such changes, both temporary and permanent, on membrane strength, interactions between the membrane and sealing material, and/or the ability of the membrane to transport hydrogen are not yet understood. However, material stability would certainly be a desired quality for any commercial or industrial applications. These results provide goals for future improvements in the technology.

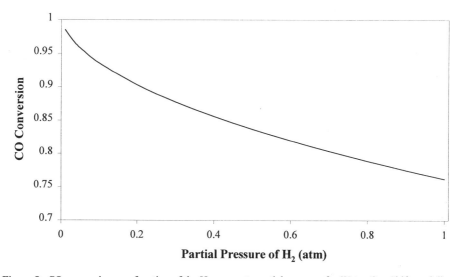

Figure 8. CO conversion as a function of the H_2 permeate partial pressure for Water-Gas Shift modeling.

High-temperature, high-pressure hydrogen flux measurements for these membranes indicated that the permeation of hydrogen was proportional to the square root of hydrogen concentration or partial pressure. The permeability of this proton-conducting membrane was found to be 9.62×10^{-3} [cm^2/min][mol/lit]$^{0.5}$, intermediate to that of palladium and iron, at 700°C. The permeability increased with temperature at 800 and 900°C. The temperature-dependence of the membrane's hydrogen permeability was expressed in an Arrhenius relationship.

A model of a tubular, plug flow membrane reactor was developed for evaluating the effect of hydrogen permeable membranes on the conversion of CO in the water-gas shift reaction. The model has been developed for very high temperature systems (>800°C) that do not employ a catalyst. Under these conditions, the presence of a membrane increased conversion of CO from 35% to 79%, with still higher values possible if hydrogen was actively removed from the permeate side of the membrane. The model can incorporate hydrogen permeation/diffusion models that are appropriate for novel high-temperature, high-pressure membranes currently being developed. This tool can be used to assess the levels of CO conversion, H_2 purity and recovery, and CO_2-rich retentate flow rate and recovery that can be realized in a reactor of specified geometry if proton-conducting membranes are incorporated. The model can also be used to provide "targets" for the hydrogen permeability required to make this technology economically feasible.

ACKNOWLEDGEMENTS AND DISCLAIMER

The hydrogen membrane testing unit at NETL was operated and maintained by R. Hirsh, J. Brannen, R. Rokicki, M. Ditillo, B. Neel, M. Schellhaas, and G. Schlata. Surface characterization data was obtained with the assistance of Edward Fisher. The work at ANL was supported by the U.S. Department of Energy – NETL, under Contract W-31-109-Eng-38. Reference in this paper to any specific commercial product, process, or service is to facilitate understanding and does not necessarily imply its endorsement or favoring by the United States Department of Energy.

REFERENCES

Adris, A.; Lim, C.; Grace, J.; The Fluidized Bed Membrane Reactor System: A Pilot-Scale Experimental Study; Chemical Engineering Science 49, No. 24B (1994) 5833-5843.

Armor, J.; Membrane Catalysis: Where Is It Now, What Needs to be Done?; Catalysis Today 25 (1995) 199-207.

Balachandran, U.; Guan, J.; Dorris, S. E.; Liu, M.; Development of Proton-Conducting Membranes for Separating hydrogen from Gas Mixtures; extended abstract of presentation at the AIChE 1998 Spring Meeting, New Orleans, LA, March 8-12, 1998.

Barbieri, G.; DiMaio, F.; Simulation of the Methane Steam reforming Process in a Catalytic Pd-Membrane Reactor; Ind. Eng. Chem. Res. 36 (1997) 2121-2127.

Barbieri, G,; Violante, V.; DiMaio, F.; Criscuoli, A.; Drioli, E.; Methane Steam reforming Analysis in a Palladium-Based Catalytic Membrane Reactor; Ind. Eng. Chem. Res. 36 (1997) 3369-3374.

Basile, A.;Criscuoli, A.; Santella, F.; Drioli, E.; Membrane Reactor for Water Gas Shift Reaction; Gas Separation and Purification 10(4) (1996a) 243-254.

Basile, A.; Drioli, E.; Santella, F.,; Violante, V.; Capannelli, G,; A Study on Catalytic Membrane Reactors for Water Gas Shift Reaction; Gas Separation & Purification 10(1) (1996b) 53-61.

Benson, H. E.; Processing of Gasification Products; Ch. 25 in Chemistry of Coal Utilization, Elliot, M., ed.; John Wiley and Sons; New York, NY, 1981.

Buxbaum, R.E.; Marker, T.L.; J; Hydrogen Transport Through Non-Porous Membranes of Palladium-Coated Niobium, Tantalum, and Vanadium; . Membr. Sci. 85 (1993) 29-38.

Cugini, A.; Krastman, D.; Lett, R.; Balsone, G.; Development of a Dispersed Iron catalysis for First Stage Coal Liquefaction; Catalysis Today 19 (1994) 395-408.

Damle, A.S.; Gangwal, S.K.; Venkataraman, V.K.; A Simple Model for a Water Gas Shift Membrane Reactor; Gas Separation & Purification 8, No. 2 (1994) 101-106.

DeRossett, A.J.; Diffusion of Hydrogen Through Palladium Membranes; Ind. Eng. Chem. 52(6) (1960) 525-528.

Enick, R.M.; Hill, J.; Cugini, A.V., Rothenberger, K.S.; McIlvried, H.G.; "A Model of a High Temperature, High Pressure Water-Gas Shift Tubular Membrane Reactor", Am. Chem. Soc., Fuel Chem. Div., Prepr. Pap., 44(4), (1999) 919-923.

Fain, D. E.; Roettger, G. E.; Coal Gas Cleaning and Purification with Inorganic Membranes; Trans. of the ASME Journal of Engineering for Gas Turbines and Power, 115 (July 1993) 628-633.

Fogler, H. S.; Elements of Chemical Reaction Engineering, Second Edition; Prentice Hall, Englewood Cliffs, NJ (1992).

Graven, W. M.; Long, F. J.; Kinetics and Mechanisms of the Two Opposing Reactions of the Equilibrium CO + H_2O = CO_2 + H_2; J. Amer. Chem. Soc. 76 (1954) 2602.

Helling, R.K.; Tester, J.W.; Oxidation Kinetics of Carbon Monoxide in Supercritical Water; Energy and Fuels, 1 (1987) 417-423.

Holgate, H.R.; Webley, P.A.; Tester, J.W.; CO in Supercritical Water: The Effects of Heat Transfer and the Water-Gas Shift Reaction on Observed Kinetics; Energy & Fuels, 6 (1992) 586-597.

Hurlbert, R.C.; Konecny, J.O.; Diffusion of Hydrogen through Palladium; J. of Chem. Phys. 34(2) (1961) 655-658.

Itoh, N.; Govind, R.; Development of a Novel Oxidative Palladium Membrane Reactor; AIChE Symposium Series 268, Govind and Itoh, eds.; New York, NY 1989.

Itoh, N.; Xu, W.; Sathe, A.; Capability of Permeate Hydrogen through Palladium-Based Membranes for Acetylene Hydrogenation; Ind. Eng. Chem. Res. 32 (1993) 2614-2619.

Keiski, R.I.; Desponds, O.; Chang, Y.; Somorjai, G.A.; Kinetics of the Water-Gas Shift Reaction Over Several Alkane Activation and Water-Gas Shift Catalysts; Applied Catalysis A: General, 101 (1993) 317-338.

Keiski, R.; Salmi, T.; Niemisto, P.; Ainassaari, J.; Pohjola, V.J.; Stationary and Transient Kinetics of the High Temperature Water-Gas Shift Reaction; Applied Catalysis A: General, 137 (1996) 349-370.

Liu, Y.; Dixon, A.; Ma, Y.; Moser, W.; Permeation of Ethylbenzene and Hydrogen Through Untreated and Catalytically Treated Alumina Membranes; Separation Science and Technology 25 (1990) 1511-1521.

Mardilovich, P.; She, Y.; Ma, Y.; Rei, M.; Defect-Free Palladium Membranes on Porous Stainless Steel Support; AIChEJ 44(2) (1998) 310-322.

Netherlands Energy Research Foundation ; Combined Cycle Project – IGCC with CO_2 Removal Project Area – An Attractive Option for CO_2 Control in IGCC Systems: Water Gas Shift with Integrated Hydrogen/Carbon Dioxide Separation (WIHYS) Process – Phase 1 Proof of Principle; Alderliesten, P. T. and Bracht, M., eds.; Contract JOU2-CT92-0158, Final Report (1995).

Parsons Infrastructure and Technology Group; Decarbonized Fuel Plants Utilizing Inorganic Membranes for Hydrogen Separation; presented at the 12th Annual Conference of Fossil Energy Materials, May 12-14, 1998, Knoxville, Tennessee.

Parsons Infrastructure and Technology Group; Decarbonized Fuel Plants for Vision 21 Applications, WYODAK Coal Substitution; DOE/FETC Report, April 13, 1999.

Peachey, N. M.; Snow, R. C.; Dye, R. C.; Composite Pd/Ta Metal Membranes for Hydrogen Separation; Journal of Membrane Sciences (in press).

Rice, S.F., Steeper, R. R.; Aiken, J. D.; Water Density Effects on Homogeneous Water-Gas Shift Reaction Kinetics; J. Phys. Chem. A, 102(16) (1998), 2673-2678.

Rostrup-Nielsen, J.R.; Catalytic Steam Reforming; Chapter 1 of Catalysis Science and Technology, Vol. 5; Anderson, J. and Boudart, M., eds; Springer-Verlag (1984), 1-118.

Rothenberger, K.S.; Cugini, A.V.; Siriwardane, R.V.; Martello, D.V.; Poston, J.A.; Fisher, E.P.; Graham, W.J.; Balachandran, U.; Dorris, S.E.; "Performance Testing of Hydrogen Transport Membranes at Elevated Temperatures and Pressures", Am. Chem. Soc., Fuel Chem. Div., Prepr. Pap., 44(4), (1999). 914-918

Singh, C.P.; Saraf, D.N.; Simulation of High-Temperature Water-Gas Shift Reactors; Ind. Eng. Chem. Process Des. Dev. 16(3) (1977)

Sun, Y.; Khang, S.; "Catalytic Membrane for Simultaneous Chemical Reaction and Separation Applied to a Dehydrogenation Reaction," Ind. Eng. Chem. Res. 27 (1988) 1136-1142.

Tsotsis, A.; Champagnie, A.; Vasileiadis, S.; Ziaka, Z.; Minet, R.; Packed Bed Catalytic Reactors; Chemical Engineering Science 47(9) (1992) 2903-2908.

Uemiya, S.; Sato, N.; Ando, H.; Kikuchi, E.; The Water Gas Shift Reaction Assisted by a Palladium Membrane Reactor; Ind. Eng. Chem. Res. 30 (1991a) 585-589.

Uemiya, S.; Sato, N.; Ando, H.; Matsuda, K.; Kikuchi, E.; Steam Reforming of methane in a Hydrogen Permeable Membrane Reactor; Applied Catalysis 67 (1991b) 223-230.

A FIRST-PRINCIPLES STUDY OF HYDROGEN DISSOLUTION IN VARIOUS METALS AND PALLADIUM-SILVER ALLOYS

Yasuharu Yokoi, Tsutomu Seki, and Isamu Yasuda

Fundamental Technology Research Laboratory
Tokyo Gas Co., Ltd.
Tokyo, 105-0023, Japan

INTRODUCTION

Steam reforming of methane has been used industrially to produce hydrogen or syngas. Recently, considerable attention has focused on an application of this reaction in the production of hydrogen for fuel cells. Among fuel cells, the polymer electrolyte fuel cell can be operated at lower temperature than other types of fuel cells. Because of this, high overall thermal efficiency would be expected if the steam reforming could be carried out at comparatively low temperatures. In fact, the steam reforming reaction is largely endothermic so that high reaction temperatures are usually required. If hydrogen is selectively removed from the reaction system, however, high reaction temperatures are not necessarily required from a thermodynamic viewpoint.

Steam reforming of methane in membrane reactors was first studied by Oertel et al.[1] They used a Pd membrane that is permeable only to hydrogen. By using the membrane reactor, the produced hydrogen can be selectively removed from the reaction system, which shifts the equilibrium of the hydrogen-producing reaction in the forward direction.[2] Therefore, higher conversions can be achieved at lower temperatures than the conventional systems. Moreover, simple and more compact system configurations can be expected.

Steam reformers equipped with the Pd membranes were developed and have been tested in Japan to produce pure hydrogen from city gas.[3] Because of the working principle of the membrane reactor, the performance of this type of steam reformer directly depends on hydrogen permeability of the membranes. This has led us to develop membranes with higher hydrogen permeability.

One of the effective approaches to increasing hydrogen permeability of the membranes is alloying. For example, the permeability of hydrogen through Pd-Ag membranes depends on the silver content and reaches a maximum at around 23% Ag by weight.[4] This maximum arises because the solubility of hydrogen into the membrane also reaches a maximum at around 23% Ag. It is very interesting that the alloying of Pd with Ag enhances the hydrogen dissolution in spite of much less solubility of hydrogen into Ag

Advances in Hydrogen Energy, edited by Padró and Lau
Kluwer Academic/Plenum Publishers, 2000

than Pd. However, no studies have ever elucidated the enhanced mechanism of the hydrogen solubility by the Ag-alloying. The key to a quantitative understanding of the mechanism is a detailed knowledge of the interaction of the dissolved hydrogen with the host metals.

In the present study, we have applied first-principles calculations to study of the interaction of the dissolved hydrogen with various metals and alloys. Our computations have focused on the heat of hydrogen dissolution, which should be directly related to the solubility of hydrogen.

METHODS OF CALCULATION

First-Principles Calculation

All calculations in this study were implemented with the CASTEP package[5], which is capable of simulating electronic structures for metals, insulators, or semiconductors. It is based on a supercell method, whereby all studies must be performed on a periodic system. Study of molecules is also possible by assuming that a molecule is put in a box and treated as a periodic system. Forces acting on atoms and stress on the unit cell can be calculated. These can be used to find the equilibrium structure.

The theoretical basis of CASTEP is the density functional theory (DFT) in the local density approximation (LDA) or gradient-corrected LDA version, as developed by Perdew and Wang (GGA).[6,7] The DFT description of electron gas interactions is known to be sufficiently accurate in most cases, and it remains the only practical way of analyzing periodic systems. LDA is known to underestimate bond lengths in molecules and cell parameters in crystals, while GGA is typically more accurate to these optimized geometries. In the present calculations, we selected GGA, which is the default setting in CASTEP.

In the present study, the valence electron orbitals were expanded in plane waves, whereas the core electrons were described using a pseudopotential concept. The pseudo-wave-function matches the all-electron wave function beyond a cutoff radius that defines the core region. Within the core region, the pseudo-wave-function has no nodes and is related to the all-electron wave function by the norm-conservation condition; that is, both wave functions carry the same charges. The conventional type of pseudopotentials can be made very accurate at the price of having to use a very high energy cutoff, as the energy cutoff needed to describe the localized valence orbitals of transition metals is extremely high. The idea of ultrasoft pseudopotentials (USP) as put forward by Vanderbilt is that the relaxation of the norm-conserving condition can be used to generate much softer potentials.[8] In this scheme the pseudo-wave-functions are allowed to be as soft as possible within the core region, so that the cutoff energy can be reduced dramatically. In the present calculations, we selected USP and an energy cut-off of 380 eV, which are set as the default for metal-hydrogen systems in CASTEP.

Models for Calculation

It has been generally accepted that hydrogen molecules dissociate into two hydrogen atoms when they dissolve into metals.[9] Figure 1 illustrates the periodic supercell models for hydrogen occupation at interstitial sites in bcc (V, Nb, Ta, Cr, Mo and W) and fcc (Ni, Pd, Pt, Cu, Ag and Au) metals. The models consist of four metal atoms and one hydrogen atom (H/M = 0.25). The initial positions of metal atoms and the initial values of lattice constants were those from experimental values of pure metals. The initial position of the hydrogen atom was either the octahedral site (O-site) or the tetrahedral site (T-site).

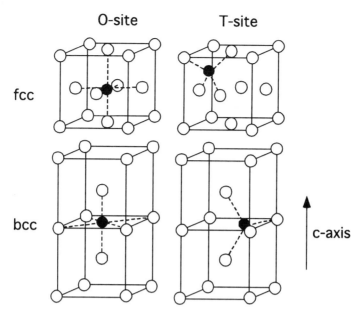

Figure 1. Periodic supercell models for hydrogen occupation at interstitial sites in fcc and bcc metals. Closed circle = hydrogen; open circles = metal.

Heat of Hydrogen Dissolution

The final geometry was obtained when the calculated forces acting on the atoms and stress on the supercell became smaller than the threshold values. To evaluate the relative expansion of the lattice induced by the interstitial hydrogen, geometry optimizations of the pure metals were also employed.

The heat of hydrogen dissolution was calculated according to the following expression,

$$E_{diss} = E_{(host + H)} - E_{host} - 1/2 \, E_{H2} \qquad (1)$$

where $E_{(host + H)}$ is the total energy of the optimized model for the hydrogen-metal system, E_{host} is of the optimized model for the pure metal, and E_{H2} is of the hydrogen molecule. The optimized bond length of H-H bond of the hydrogen molecule was confirmed to be nearly equal to the experimental value.

RESULTS

Hydrogen Dissolution in Pure Metals

Lattice constants. From the geometry optimization for the pure metals, all optimized fcc and bcc cells were nearly cubic. The lattice constants of the optimized cells are plotted versus the experimental values in Figure 2. The O-site occupation for the bcc metals was confirmed not to be a stationary point on the potential energy surface except for V and Mo, whereas the T-site occupation was found to be the stationary point. As for the fcc metals, both interstitial sites were confirmed as the stationary point.

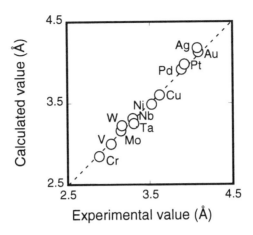

Figure 2. Correlation between calculations and experiments for lattice constants of various metals.

Lattice expansion. All optimized supercells for the hydrogen-metal systems were found to expand in comparison with pure metals. The optimized fcc cells for both the O- and T-site occupation maintained cubic. The average percentage of the lattice expansion for the O- and T-site occupation was 1.5% and 2.5%, respectively. The optimized bcc cells, on the other hand, were distorted. The average percentage for the O- and T-site occupation was 1.5% and 1.7%, respectively. Figure 3 shows the changes in the lattice constants for V as the representative of the bcc metals. The hydrogen occupation at the O-site in V was found to induce the lattice expansion of about 4.7% in the direction of the c-axis, in spite of small contraction of the lattice in the directions of the a- and b-axis. Figure 4 shows the changes in the lattice constants for Pd as the representative of the fcc metals.

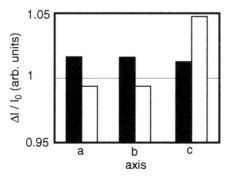

Figure 3. Relative expansion of the lattice constants of V induced by interstitial hydrogen. Open bars = O-site occupation; closed bars = T-site occupation.

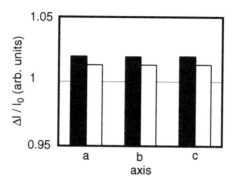

Figure 4. Relative expansion of the lattice constants of Pd induced by interstitial hydrogen. Open bars = O-site occupation; closed bars = T-site occupation.

Changes in Volume of Pd by Hydrogen Dissolution. In the case of Pd, the relationship between the lattice expansion and the hydrogen concentration was also examined. The models used in this examination were constructed by adding hydrogen atoms 'one by one' to the original supercell model. The volume of the optimized fcc cells (in Å^3) was 59.82, 62.17, 64.29, 66.04 and 67.82 for H/Pd ratio of 0, 0.25, 0.5, 0.75 and 1, respectively.

Heat of Hydrogen Dissolution. Figure 5 shows E_{diss} for the fcc metals. As for Ni and Pd, the O-site occupation was found to be more stable than the T-site occupation, whereas the T-site occupation was found to be more stable for Pt, Cu, Ag and Au.

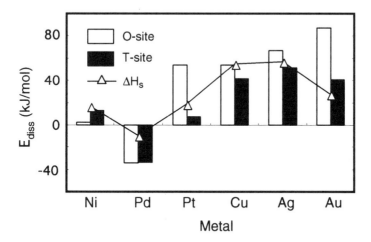

Figure 5. Calculated E_{diss} for various fcc metals in comparison with experimental ΔH_s.

Figure 6 shows E_{diss} for the bcc metals. The T-site was suggested to be the stable interstitial hydrogen site for all bcc metals evaluated in this study.

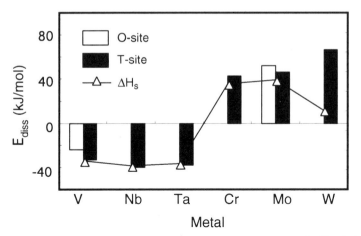

Figure 6. Calculated E_{diss} for various bcc metals in comparison with experimental ΔH_s. The O-site occupation is not stable for Nb, Ta, Cr, and W.

(a) O_α-site (b) O_β-site

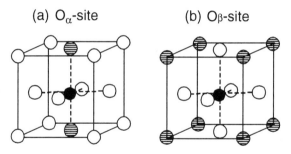

Figure 7. Periodic supercell models for hydrogen occupation at two distinguishable O-sites in the Pd/Ag alloy. (a) O_α-site; (b) O_β-site. Closed circles = H; open circles = Pd; shaded circles = Ag.

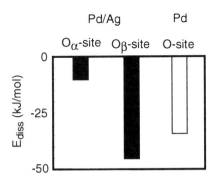

Figure 8. Calculated E_{diss} for Pd/Ag alloys in comparison with that for pure Pd.

Hydrogen Dissolution in Pd/Ag Alloys

To study the effects of alloying of Pd with Ag on the heat of hydrogen dissolution, E_{diss} for the Pd/Ag alloy was evaluated in the same manner. Figure 7 illustrates the periodic supercell models for the Pd/Ag alloy. These models consist of three Pd atoms, one Ag atom, and one hydrogen atom. There are two distinguishable O-sites in these models; one is the center of the octahedron consisting of four Pd atoms and two Ag atoms (O_α-site), and the other is the center of the octahedron consisting of six Pd atoms (O_β-site).

The calculated E_{diss} is shown in Figure 8 in comparison with that of the O-site occupation for the pure Pd. The value of $\mid - E_{diss} \mid$ for the O_β-site occupation was larger, whereas that of the O_α-site occupation was smaller, than that of the O-site occupation for the pure Pd.

DISCUSSION

Lattice Expansion of Pure Metals by Hydrogen Dissolution

As for the optimized lattice constants for the pure metals and the optimized bond length of H-H bond of hydrogen molecule, the agreement between the calculations and the experiments is quite satisfactory. This suggests that the calculated values of the lattice constants of the hydrogen-metal systems are reliable. In this study, we have estimated the lattice expansion of pure Pd induced by the hydrogen occupation at the interstitial sites; for example, a H/Pd ratio of 0.5 results in an expansion of 7.5% by volume. An experimental value corresponding to this estimation has been reported to be about 10% in the literature.[4] We have also evaluated the lattice expansion of V. From the result shown in Figure 3, the O-site occupation in V was found to induce the lattice expansion of 4.7% in the direction of the c-axis. This is qualitatively in line with the experimental observation that the lattice is expanded in the direction of the c-axis by about 10%.[9]

Hydrogen Sites in Pure Metals

Figure 5 shows that the hydrogen occupation at both interstitial sites is stable for the fcc metals. For Ni and Pd, the O-site occupation is suggested to be more stable than the T-site occupation. As has been expected from the results, the O-site occupation for Ni and Pd has been experimentally observed.[9,10] As for the other fcc metals (Pt, Cu, Ag and Au), the T-site occupation is more stable than the O-site occupation. This suggests that the T-site occupation in these fcc metals should be experimentally observed, although there is no reliable data to confirm it.

On the other hand, the T-site occupation is stable for the bcc metals shown in Figure 6. As has been expected from the results, the preference of T-site occupation has already been reported for these bcc metals.[9,10] As for V, the O-site occupation, which occurred with the expansion of about 10% in the direction of the c-axis, has been also reported.[11] This also agrees with the present calculation result that the O-site occupation in V is stable as well as the T-site occupation (shown in Figure 6). The O-site occupation induces the lattice expansion with the distortion, as shown in Figure 3.

Heat of Hydrogen Dissolution for Pure Metals

As the experimental heat of hydrogen dissolution (ΔH_s),[12] plotted on Figure 5 and Figure 6, was obtained by applying Sieverts's law, the values may have some errors in high hydrogen concentration (far from H/M = 0) ranges. Nevertheless the calculated E_{diss} correlated well with the experimental ΔH_s.

Before turning to a closer examination of the alloying of Pd with Ag, it is desirable to discuss the deviation of the calculated E_{diss} from the experimental ΔH_s for Pd. The deviation for Pd shown in Figure 5 is somewhat larger than that of V, Nb and Ta, which are exothermic for the hydrogen dissolution, as is Pd. The dependence of ΔH_s for Pd on hydrogen concentration has been experimentally studied in the previous literature.[9] In this literature, the experimental heat of hydrogen dissolution for Pd at H/Pd = 0.25, which is equal to the H/M ratio of the calculated model shown in Figure 1, have been reported as below –20 kJ/mol. This suggests that the deviation of E_{diss} from the experimental value is reduced by the correction for the hydrogen concentration. The calculated E_{diss} is believed to be reliable enough to discuss the qualitative difference in the heat of hydrogen dissolution between the pure Pd and the Pd/Ag alloys.

Heat of Hydrogen Dissolution for Pd/Ag Alloys

The hydrogen permeability of metals is proportional to their solubility and diffusivity of hydrogen. Since the hydrogen solubility increases with $-\Delta H_s$, increasing of $-\Delta H_s$ is an effective means of improving hydrogen permeability. It has been reported that the high hydrogen solubility in Pd/Ag alloys leads to high permeability in the 20 to 25% Ag ranges.[4] According to this literature,[4] about 150 mg of hydrogen can dissolve in 100g of the Pd/Ag alloys at 1 atmosphere and 183°C. This H/M ratio is below 0.25. As the hydrogen dissolution in Pd/Ag alloys is exothermic, the H/M ratio decreases with the increase in temperature. The steam reformers are usually operated above 500°C,[3] and the hydrogen diffusion membranes made of the Pd/Ag alloys have hydrogen of the H/M ratio below 0.25. The simple calculation models for 25% Ag - Pd shown in Figure 7 have three O_α-sites and one O_β-site. Since the O_β-site occupation has been found to be more stable than the O_α-site occupation in the present study, hydrogen is suggested to occupy only the O_β-sites below 0.25 of the H/M ratio. As shown in Figure 8, The value of | - E_{diss} | for the O_β-site occupation is larger than that for the O-site occupation in the pure Pd. In this way, it suggests that the $-\Delta H_s$ of the Pd/Ag alloys is greater than that of the pure Pd in the low hydrogen concentration ranges. This is the reason for the high hydrogen solubility in the Pd/Ag alloys at higher temperatures and contributes to high hydrogen permeability.

Problems and Perspective

Our goal is to develop membranes with higher hydrogen permeability than the conventional Pd/Ag alloys and with mechanical reliability under practical operating conditions.

As mentioned above, the hydrogen permeability of metals and alloys depends not only on their solubility but also on their diffusivity. This study only focused on the former because of the limitation of our computational resources. As for the Pd/Ag alloys, the diffusivity tends to decrease with the increase in Ag content.[13] The significant increase in the solubility by the Ag-alloying, which is discussed in the present study, has led to enhanced hydrogen permeability in spite of the decrease in the diffusivity with Ag content. In the next step, the calculation of the diffusivity will be needed for more precise estimation of the hydrogen permeability of metals and alloys.

The models used in this study were limited under the periodic boundary condition so that the models should be applied to single-phase materials, although the formation of the β phase was experimentally observed to coexist with the α phase at temperatures below

300°C and pressures below 20 atmosphere in Pd-H system.[14] The β phase has a considerably expanded lattice compared with the α phase; for example, a H/Pd ratio of 0.5 results in an expansion of about 10% by volume,[4] which should cause mechanical damage to the membranes after repeated dissolution/evolution cycles during the operation of the reformer. The Ag-alloying is effective in depressing the α-β miscibility gap to well below room temperature as well as in enhancing the hydrogen permeability. The hydrogen-induced lattice expansion of the single-phase metals has been theoretically verified in the present study. This will be a basis for the theoretical evaluation of the mechanical reliability of the hydrogen separation membranes.

CONCLUSIONS

Interactions between various metals and hydrogen have been studied by using a periodic density functional theory at a generalized gradient approximation. Geometry optimizations were carried out for lattice constants of various metal-hydrogen systems as well as for stable sites of interstitial hydrogen. Both tetrahedral site (4-coordinated sites) hydrogen and octahedral site 6-coorinated sites) hydrogen was found to be stable and to induce lattice expansion. Reported experimental heat of hydrogen dissolution has been well reproduced from the calculated total energies. Similar estimations were carried out for Pd/Ag alloys that are used as hydrogen permeable membranes in hydrogen production. The theoretical calculations suggest that alloying of Pd with Ag enhances hydrogen dissolution in Pd. These can explain fairly well the increased hydrogen permeability by alloying Pd with Ag. The first-principles approach can be a powerful tool for the design of novel alloys for hydrogen separation membrane.

REFERENCES

1. M. Oertel, J. Schmitz, W. Weirich, D. Jendryssek-Neumann and R. Schulten, Chem. Eng. Technol., 10 (1987) 248.
2. E. Kikuchi, CATTECH, (1997) 67.
3. Y. Ohta, M. Gondaira, K. Kobayashi, Y. Fujimoto and K. Kuroda, in Extended Abstracts of AIChE Annual Meeting, 1994, San Francisco.
4. A. G. Knapton, Platinum Metals Review, 21 (1977) 44.
5. The CASTEP code, originally written by M. C. Payne, is marketed by Molecular Simulations, Inc.
6. J.P. Perdew and Y. Wang, Phys. Rev. B45 (1992) 13244.
7. J.A. White and D.M. Bird, Phys. Rev. B50 (1994) 4954.
8. D. Vanderbilt, Rhys. Rev. B41 (1990) 7892.
9. Y. Fukai, *"The Metal Hydrogen System"*, Springer, Series in Materials Science Vol. 21 (1993) Chap. 4.
10. D. Richter, R. Hempelmann and R.C. Bowman, *"Hydrogen in Intermetallic Compounds II : Surface and Dynamic Properties,Applications"*, Springer, Topics in Applied Physics, Vol. 67 (1992) Chap. 3.
11. I. Okuda, H. Asano and M. Hirabayashi, Trans. JIM, 21 (1980) 89.
12. H.M. Lee, Metall. Trans., 7A (1976) 431.
13. E. Wicke and H. Brodowsky, *"Hydrogen in Metals II: Application-oriented Properties"*, Springer-Verlag, Topics in Applied Physics, Vol. 29 (1978) Chap. 3.
14. F.A. Lewis, *"The Palladium Hydrogen System"*, Academic Press, 1967.

INVESTIGATION OF A NOVEL METAL HYDRIDE ELECTRODE FOR Ni-MH BATTERIES

N.Rajalakshmi[1], K.S.Dhathathreyan[1], and Sundara Ramaprabhu[2]

[1]Centre for Electrochemical and Energy Research
SPIC Science Foundation
111, Mount Road
Guindy
Chennai 600 032, INDIA

[2]Magnetism and Magnetic Materials Laboratory
Department of Physics
Indian Institute of Technology
Chennai 600 036, INDIA

ABSTRACT

The science and technology of a nickel metal hydride battery that stores hydrogen in the solid hydride phase has many advantages, including high energy density, high power, long life, tolerance to abuse, wide range of operating temperatures, quick-charge capability, and totally sealed and absolutely maintenance-free operation. The primary goal of metal hydride research for use in battery electrodes has been a cell capacity higher than that presently available with conventional Ni-Cd technology. A novel metal hydride consisting of Y, Zr, Mn, Fe, Co, V and Cr: $Y_xZr_{1-x}Mn_mFe_nCo_pV_oCr_q$ (m+n+o+p+q=2) is shown to give a higher electrochemical capacity. The electrochemical properties, such as electrode potential, reversible electrochemical capacity and diffusion coefficient as a function of state of charge in electrodes, are investigated in order to evaluate their suitability of the material as an electrode. The reversible electrochemical capacity of the electrode is found to be in excess of 450 mAh/g. Hydrogen concentration is estimated as $r = n_H/n_{f.u.} = 3.5$. The process that occurs in the electrode during charge and discharge has been studied by cyclic voltammogram (CV) experiments, carried out at different sweep rates. It is found that at low sweep rates, the hydrogen concentration on the surface increases due to longer polarization and approaches a value that favors a metal hydride formation. The diffusion coefficients are also evaluated with respect to state of charge.

Advances in Hydrogen Energy, edited by Padró and Lau
Kluwer Academic/Plenum Publishers, 2000

INTRODUCTION

The need for high energy density storage batteries has been growing in recent years. The advent of cellular telephones, portable computers and new cordless appliances and tools has made this need even more urgent. Although conventional storage batteries like Ni-Cd and lead acid batteries have been further improved in design and packaging in recent years, there is still a need for improved performance and power densities. The innate toxicity of cadmium and lead also has come under scrutiny. The use of metal hydrides as active negative electrode material in rechargeable alkaline batteries has been studied recently. The basic objective of this research has been to get higher energy density compared to Ni-Cd. However, many problems in areas such as cycle life, rate capability, charging efficiency, manufacturing process and cost have hindered application of conventional hydride materials to commercial sealed cells[1].

The diverse properties required for a superior MH battery electrode can be attained by the engineering of new hydrogen storage materials on the basis of the concepts of structural and compositional disorder. The hydrogen storage properties of the alloys, whether as an electrode or as a storage medium, are related to their composition and vary widely with small changes in the type and the amount of substituent elements[2]. Thus, depending on the battery use, it is possible to design the electrode characteristics by changing the alloy compositions. The choice of the material is based on the desirable metal-hydrogen bond strength for use as electrodes in aqueous media. This approach allows one to consider a range of alloys for electrode materials containing elements that, if used alone, would be unacceptable for thermodynamic reactions, in particular oxidation or corrosion. Certain elements like Mg, Ti, V, Zr, Nb and La can increase the number of hydrogen atoms stored per metal atom, whereas elements like V, Mn and Zr allow the adjustment of the metal-hydrogen bond strength. Catalytic properties to ensure sufficient charge and discharge reaction rates and gas recombination can be provided by elements like Al, Mn, Fe, Co and Ni, whereas desirable surface properties such as oxidation and corrosion resistance, improved porosity, and electronic and ionic conductivities are provided by elements like Cr, Mo and W. The wide range of physical and chemical properties that can be produced in these alloys allows the MH battery performance to be optimized.

Two types of fundamental metal hydride electrodes comprising the AB_5 and AB_2 classes of alloys are currently of interest. The AB_5 alloys with A = rare earth or mischmetal, B = Ni and/or other transition metal are investigated. $LaNi_5$ has been well-investigated because of its utility in conventional hydrogen storage applications, but it is very expensive and corrodes rapidly. The commercial AB_5 electrodes use mischmetal, a low-cost combination of rare earth elements, as a substitute for La. The partial substitution of Ni by Co, Ce, Mn and Al increases the thermodynamic stability of the hydride phase, the corrosion resistance and hence the cycle life. However, the substitution reduces the hydrogen storage capacity.

The AB_2 or Laves phase alloy electrodes are multiphase alloys, where the composition is adjusted to provide one or more hydride-forming phases. The surface is the corrosion resistant due to the formation of semi-passivating oxide layers. There are few systematic guidelines to predict the alloy behavior. Hence, the AB_2 alloy electrodes may be more attractive than AB_5 electrodes in terms of cost and energy density. AB_2 hydrogen storage alloys are promising for negative electrodes in Ni-MH batteries for their higher discharge capacities and better resistance to oxidation than those of AB_5 alloys[3,4]. As AB_2 type multiphase hydrogen storage alloys are mainly composed of C14, C15 Laves phases and of some solid solutions with bcc structure, there exist abundant boundaries, with enriched electrochemically catalytic elements as active reaction sites and diffusion pipes for transporting reactants and products. However, the activation and electro-catalytic activity of the alloy electrode is still inferior to that of the MmNi$_5$ based alloy[5]. The main reasons for the slow activation and low electro-catalytic activity are due to the formation of an oxide layer on the surface of the alloy grains. This oxide layer is a bad electrical conductor, hence impedes the diffusion of hydrogen and the absence of Ni content layer on these AB$_2$ alloy surface results in a lower surface activity for hydrogen absorption/desorption[6]. Partial substitution for both the A and B sites, dramatically improves the cycle lifetime and the electrochemical discharge capacity has been optimized by substituting various metals as partial constituents and by moving the composition ratio from stoichiometry. This has resulted in the preparation of four- or five-component Laves phase hydrogen storage alloys with discharge capacities over 400 mAh/g[7]. Laves phase alloys of composition (Zr,Ti)(Ni,Mn,M)$_x$ where M = Cr,V,Co,Al and x = 1.9 to 2.1 with hexagonal C14 or cubic C15 structure have been studied in order to select the most suitable AB$_2$ alloys as an active material for Ni-MH batteries[8]. The study of hydrogen storage performance and electrochemical properties of ZrMn$_{1-x}$V$_x$Ni$_{1.4+y}$ (x= 0.5,0.7; y = 0.0 to 0.6) revealed that the major factor controlling the electrode properties was the specific reaction surface area and the exchange current density depending upon the composition. The maximum electrochemical capacity experimentally observed was 338 mAh/g for ZrMn$_{0.5}$V$_{0.5}$Ni$_{1.4}$, and the discharge efficiency was found to be 85%. Hence the fundamental studies on the charge/discharge characteristics are required for improving the performance of a Ni-MH battery[9,10]. The present work focuses on elucidating the electrochemical properties on AB$_2$ alloy with the composition Y$_x$Zr$_{1-x}$Mn$_m$Fe$_n$Co$_p$V$_o$Cr$_q$ (m+n+o+p+q=2) and the processes that occur during charge and discharge and the rate of diffusion of hydrogen in the electrode has also been presented.

EXPERIMENTAL

Alloys of composition Y$_x$Zr$_{1-x}$Mn$_m$Fe$_n$Co$_p$V$_o$Cr$_q$ (m+n+o+p+q=2) are prepared by arc melting the constituent pure elements in an arc furnace under argon atmosphere of 0.5 bar. An excess of Mn (6%) is added to the total mixture following the normal procedure since the vapor pressure of Mn is very low. To reach homogeneity, the samples are re-melted several times, turning them upside down after each solidification. During the preparation, the weight loss of the specimens is less than 0.7% for all the samples. The resulting buttons are sealed in quartz ampoules and annealed in vacuum at 10^{-6} torr at 1173 K for 48 h and slowly cooled to room temperature. The alloys thus prepared are silver white and brittle.

Powder X-ray diffractograms of the alloys are taken using Fe K$_\alpha$ radiation. The X-ray powder patterns are indexed on the basis of hexagonal C14 Laves phase structure. The lattice constants are evaluated by least square refinement.

For the electrochemical measurements, the electrodes are prepared by grinding the alloy to 75 μm and mixing it with copper powder in the ratio of 1:3 with a PTFE binder. The putty form of the mixture is mechanically pressed onto a current collector (Ni mesh) at room temperature (RT). The electrode is then sintered at 300°C for about 3 hrs under vacuum. The geometric area of the electrode is about 2 cm^2. The counter electrode is Pt and it has a much larger geometric area than the metal hydride electrode. The electrolyte is 31% KOH, as used in the alkaline batteries, and is prepared from reagent-grade KOH and deionized water. The electrode is tested for its charge-discharge characteristics, initial capacity, cycle life and diffusion coefficient. The electrochemical measurements are carried out in a flooded electrolyte condition in open cells. Potentials are monitored using a saturated calomel as the reference electrode. The electrochemical measurements include the diffusion coefficient as a function of state of charge and also the electrochemical processes that occur during charge/discharge. These measurements are carried out using an EG&G galvanostat/potentiostat Model 273.

RESULTS AND DISCUSSION

The powder X-ray diffractograms of $Y_xZr_{1-x}Mn_mFe_nCo_pV_oCr_q$ $(m+n+o+p+q=2)$ alloys taken at RT using Fe $K\alpha$ radiation confirm the formation of single phase C14 hexagonal structure. The lattice constants and the unit cell volume are evaluated using a least square refinement technique (Table 1). The unit cell constants and volume decrease with a partial substitution of Fe by Co according to a $(\text{Å}) = 5.01 - 0.0538\ x_{Co}$; $c(\text{Å}) = 8.27 - 0.255\ x_{Co}$ and $V(\text{Å}^3) = 180 - 9.35\ x_{Co}$ as expected because of the partial replacement of Fe with less metallic radius metal Co. Figure 1 shows that there is a linear decrease in the cell constants and volume with increase of Co content.

Table 1. The unit cell constants and volume of $Y_xZr_{1-x}Mn_mFe_nCo_pV_oCr_q$ $(m+n+o+p+q=2)$ alloys

Alloy	a (Å)	c (Å)	V (Å³)
$Y_xZr_{1-x}Mn_mFe_{0.8}Co_{0.2}V_pCr_q$	5.006	8.200	177.9
$Y_xZr_{1-x}Mn_mFe_{0.7}Co_{0.7}V_pCr_q$	4.996	8.172	
$Y_xZr_{1-x}Mn_mFe_{0.5}Co_{0.5}V_pCr_q$	4.985	8.155	175.5

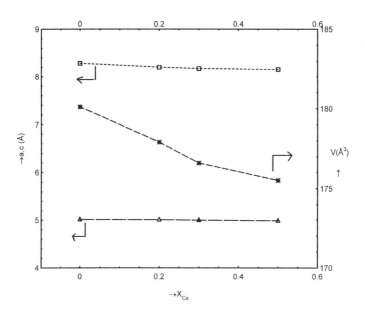

Figure 1. The dependence of unit cell constants and volume of $Y_xZr_{1-x}Mn_mFe_nCo_pV_oCr_q$ $(m+n+o+p+q=2)$ alloys.

The MH materials used for a Ni/MH battery electrode must satisfy an extensive list of requirements. The electrochemical capacity of a hydride electrode depends on the amount of reversibly absorbed hydrogen in the hydriding material, and consequently, the energy storage capacity of the battery. It is desirable to have a high electrode storage capacity that is electrochemically reversible.

Figure 2 shows the charge/discharge curves of the electrode after 30 cycles. The electrode shows the highest electrochemical charging capacity of 470 mAh/g, which corresponds to a hydrogen concentration of 3.5 hydrogen atoms/formula unit. The charging and discharging potentials are around –0.9V and –0.5V, respectively, and the hydrogen evolution is found to occur around –1.35 V.

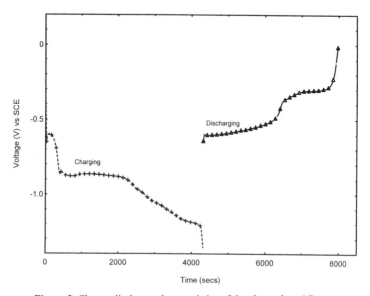

Figure 2. Charge-discharge characteristics of the electrode at 1C rate.

This shows that the metal-hydrogen bond is not weak, and hence hydrogen can react with the alloy for hydride formation. However, if the bond is very strong, the metal hydride electrode is extensively oxidized, and cannot store hydrogen reversibly. The coulombic efficiency of the electrode is found to be 85%. The electrochemical charging capacity is found to be slightly lower when the electrode is charged at the 2C rate (80 mA) compared to the 1C rate (40 mA) as shown in Figure 3.

The Ni/MH battery operates in a strongly oxidizing medium composed of a high concentration alkaline electrolyte. Because many chemical elements react to form hydrides in an alkaline electrolyte, it follows that if these elements are used as electrodes, they will oxidize and fail to store hydrogen reversibly. As already mentioned, the highest electrochemical capacity is observed only after 30 cycles and then the capacity remained constant for almost 100 cycles, as shown in Figure 4. This clearly indicates that degradation of the electrode material is very slow.

Another consideration in the use of hydride materials in Ni/MH batteries is related to the electrochemical kinetics and transport processes. The power output of the battery depends critically on these processes. During discharge, hydrogen stored in the bulk metal must be brought to the electrode surface by diffusion. The hydrogen then must react with hydroxyl ions at the metal electrolyte interface. As a consequence, surface properties such as oxide thickness, electrical conductivity, surface area, porosity and the degree of catalytic activity affect the rate at which energy can be stored in and removed from the Ni/MH battery. In order to estimate the diffusion coefficient parameter D/a^2, the following procedure is carried out. The alloy is fully activated by charging and discharging the electrode for about 30 cycles. Then the electrodes with different states of charge are equilibrated until an equilibrium potential is reached. Initially, the activated electrode is discharged at constant potential at three different states of charge. In order to secure a zero concentration of hydrogen at the surface of each individual particle, the electrode is discharged at a constant anodic potential of –0.56V (SCE). Discharge curves obtained for seven different charged states are given in Figure 5.

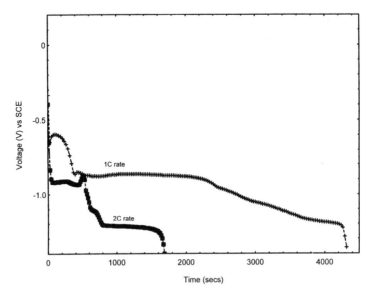

Figure 3. Capacity dependence with different charging rates.

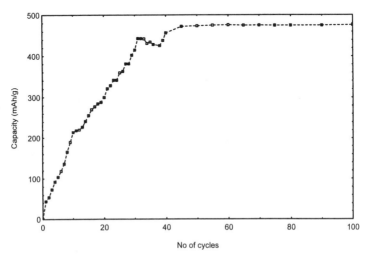

Figure 4. Capacity variation with Cycle Number.

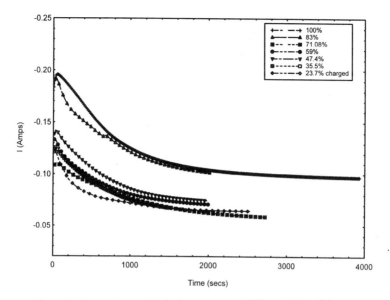

Figure 5. Constant potential discharge curves at different states of charge.

For large times in Figure 5, an approximately linear relationship exists between log (i) and t, which is consistent with the equation

$$\log(i) = \pm\log\left[6FD\frac{(C_0 - C_s)}{da^2}\right] - \frac{\Pi^2}{2.303}\frac{D}{a^2}t \qquad (1)$$

where C_0 is the concentration in the bulk of the alloy and a constant surface concentration C_s, $\pm$ sign indicates the charge (-) and discharge (+) process[11]. From the slope of the linear portion of the corresponding curves in Figure 4, the ratios of D/a^2 are estimated.

The diffusion coefficient can also be determined from the amount of charge transferred[12]. Assuming that the hydride alloy particles are in spherical form, the diffusion equation is

$$\frac{\partial(rc)}{\partial t} = D\frac{\partial^2(rc)}{\partial t^2} \qquad (2)$$

where c is the concentration of hydrogen in the alloy, t is the time, D is an average (or integral) diffusion coefficient of hydrogen over a defined concentration range, and r is a distance from the center of the sphere. Since the discharge process is carried out under a constant current condition, it is reasonable to assume a constant flux of the species at the surface and uniform initial concentration of hydrogen in the bulk of the alloy. Thus, the value of D/a^2 may be evaluated for large transition times, τ,

$$\frac{D}{a^2} = \frac{1}{15(Q_0/I - \tau)} \qquad (3)$$

where Q_0 is the initial specific capacity (c/g), I is the current density (A/g) and τ is the transient time (s), that is, the time required for the hydrogen surface concentration to become zero. The ratio Q_o/I corresponds to the discharge time necessary to discharge completely the electrode under hypothetical conditions, when the process proceeds without the interference of diffusion. D/a^2 are calculated using the above equation for different states of charge and are given in Figure 6.

As shown in Figure 6, the ratio of D/a^2 is almost independent of the state of charge with an average value of 0.37×10^{-5} s^{-1}. Assuming that the average particle radius is 15µm, the effective diffusion coefficient of hydrogen through the $Y_xZr_{1-x}Mn_mFe_nCo_pV_oCr_q$ $(m+n+o+p+q=2)$ electrode is estimated to be 8.32×10^{-12} cm^2/s. This value is consistent with the diffusion coefficient of hydrogen in nickel alloys, which is in the range 20×10^{-12} to 56×10^{-12} cm^2/s, estimated using a permeation technique.[13-15]

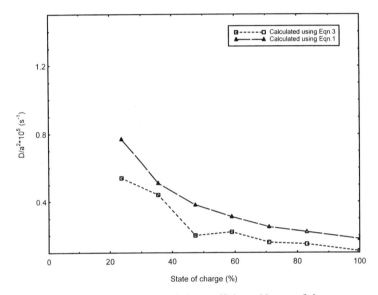

Figure 6. Dependence of diffusion coefficient with state of charge.

For the battery to be operated as a sealed system, it must also tolerate the consequences of chemical reactions that occur during cell overcharge and over-discharge. In overcharge, oxygen gas is generated in the $Ni(OH)_2$ positive electrode and must recombine with hydrogen at the MH electrode to form water. In over-discharge, hydrogen generated at the positive electrode must recombine with oxygen at the surface of the MH electrode to form water. These gas recombination reactions must occur at sufficiently high rates to avoid pressure build up. Figure 7 shows the cyclic voltammogram (CV) of an activated sample of the alloy after 5 cycles ranging from –0.6 to –1.4 V vs SCE at various scanning rates. The definite peaks at potentials ranging from –0.9 to –1.2 V are observed and these absorption peaks indicate that there are H atoms absorbed by the metal due to the formation of the hydride.

The counterparts responsible for the desorption of hydrogen from the alloys are also observed and are shown in Figure 8 in the range –0.1 to –0.6. At low sweep rates, the hydrogen concentration on the surface increases due to longer polarization. The hydrogen concentration approaches a value for the hydride formation. The formation of hydrogen absorption and desorption peaks even at higher sweep rates clearly indicating that the gas recombination reactions are fast in these electrodes. This, in turn, shows that the catalytic activity at the MH electrode is sufficient to promote the dissociation of hydrogen.

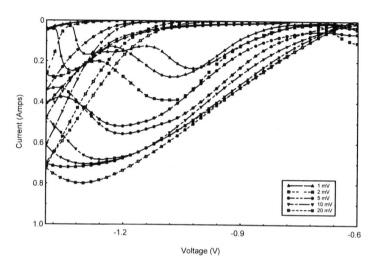

Figure 7. Cyclic voltammogram (CV) of the electrode for absorption at various sweep rates.

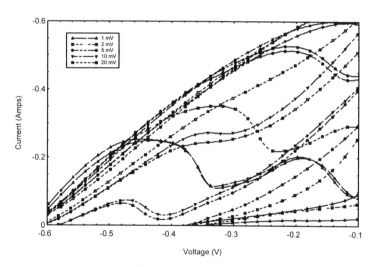

Figure 8. Cyclic voltammogram (CV) of the electrode for desorption at various sweep rates.

CONCLUSION

The reversible electrochemical capacity of the electrode prepared with the alloy $Y_xZr_{1-x}Mn_mFe_nCo_pV_oCr_q$ (m+n+o+p+q=2) is found to be in excess of 450 mAh/g and hydrogen intake capacity is estimated as r (= $n_H/n_{f.u.}$) = 3.5. Charging and discharging carried out at different sweep rates show that at low sweep rates, the hydrogen concentration on the surface increases due to longer polarization, with the concentration of hydrogen approaches a value that favors a metal hydride formation.

ACKNOWLEDGEMENTS

The authors are thankful to the Management, SPIC Science Foundation, for the financial assistance and for the facilities provided to carry out this work.

REFERENCES

1. S.R. Ovshinsky, M.A. Fetcenko, J. Ross, Science, 260 (91), 176 (1990)
2. T. Sakai, K. Oguro, H. Miyamura, N. Kuriyama, A. Kato, H. Ishikawa and K. Iwakura, J. Less Common Metals 231, 645(1995)
3. X.L. Wang, S. Suda, S. Wakao, Z. Phys. Chem 183, 297(1994)
4. H. Sawa, S. Wakao, Mater. Trans JIM, 31, 487(1990)
5. A. Zuttel, F. Meli and L. Schlapbach, J. Alloys and Compounds 231, 645(1995)
6. H. Nakano, I. Wada and S. Wakao, Denki Kagaku, 672, 953(1994)
7. B. Knosp, C. Jordy, Ph. Blanchard and T. Berlureau, J. Electrochem Soc 145, 1478(1998)
8. Dong-Myung Kim, Sang-Min Lee, Jae-Han Jung, Kuk-Jin Jang and Jai-Young Lee, J. Electrochem Soc 145, 93(1998)
9. H. Inoue, K. Yamataka, Y. Fukumoto and C. Iwakura, J. Electrochem Soc 143, 2527(1996)
10. M.P. Sridhar Kumar, W. Zhang, K. Petrov, A.A. Rostami and S. Srinivasan, J. Electrochem Soc 142, 3424(1995)
11. G. Zheng, B.N. Popov and R.E White, J. Electrochem Soc 142, 2695(1995)
12. G. Zheng, B.N. Popov and R.E White, J. Electrochem Soc 143, 834(1996)
13. W.M. Robertson, Metall Trans., 8A, 1709(1977)
14. R.M. Latanision and M. Kurkela, Corrosion, 39, 174(1983)
15. W.M. Robertson, Metall Trans., 10A, 489(1979)

HYDROGEN STORAGE USING SLURRIES OF CHEMICAL HYDRIDES

Andrew W. McClaine, Ronald W. Breault, Jonathan Rolfe,
Christopher Larsen, Ravi Konduri, Gabor Miskolczy, and Frederick Becker

Thermo Power Corporation
45 First Avenue
Waltham, MA 02154-9046

INTRODUCTION

The usual storage technologies considered for hydrogen are compressed hydrogen, liquid hydrogen, metal hydrides, and carbon-based storage systems. Over the past couple of years another method of hydrogen storage and transmission has been under investigation that offers some significant advantages over the usual hydrogen storage technologies. Thermo Power Corporation has been developing a chemical hydride slurry approach. In this approach, a light metal hydride is used as the hydrogen carrier and storage media. Light metal hydrides such as lithium hydride, magnesium hydride, sodium hydride, and calcium hydride produce hydrogen when they react with water. These materials are typically dry solids at ambient conditions. The oil in the slurry protects the hydride from unintentional contact with moisture in the air and makes the hydride pumpable. At the point of storage and use, a chemical hydride/water reaction is used to produce high-purity hydrogen. An essential feature of this approach is the recovery and recycle of the spent hydride at centralized processing plants to produce new hydride slurry, resulting in an overall low cost for hydrogen. This chemical hydride slurry system has several benefits:

- it greatly improves the energy transmission and storage characteristics of hydrogen as a fuel,
- it provides a hydrogen storage medium that is stable at normal environmental temperatures and pressures,
- it is pumpable and easily transported,
- it has a high gravimetric and volumetric energy density,
- with the use of a properly designed reactor it can provide hydrogen at elevated pressures without the use of a compressor,
- it produces the hydrogen carrier efficiently and economically from a low cost carbon source, and
- since the production of the hydride is a carbo-thermal process performed at a centralized plant, CO_2 resulting from the carbo-thermal process for refining lithium is concentrated and amenable to sequestration.

Advances in Hydrogen Energy, edited by Padró and Lau
Kluwer Academic/Plenum Publishers, 2000

Our preliminary economic analysis of the chemical hydride slurry cycle indicates that hydrogen can be produced for $3.65/GJ ($3.85/MMBtu) based on a carbon cost of $1.35/GJ ($1.42/MMBtu) and a plant sized to serve a million cars per day. This compares to current costs of approximately $5.75/GJ ($6.06/MMBtu) to produce hydrogen from $2.85/GJ ($3.00/MMBtu) natural gas, and $24.50/GJ ($25.83/MMBtu) to produce hydrogen by electrolysis from $0.05 per kWh electricity. The present standard for production of hydrogen from renewable energy is photovoltaic-electrolysis at $41.80/GJ ($44.07/MMBtu). Costs have been derived from Padró (1999).

Two programs are currently under contract. These programs are sponsored by the U.S. Department of Energy, the California South Coast Air Quality Management District, the Southern Illinois University, and Thermo Power Corporation. The objective of the first of these two programs is to investigate the technical feasibility and economic viability of the metal hydride organic slurry approach for transmission and storage of hydrogen with analysis and laboratory-scale experiments, and to demonstrate the critical steps in the process with bench-scale equipment. The objective of the second of these two programs is to design and fabricate a 50 kW electric-power-equivalent hydrogen supply system utilizing chemical hydride organic slurry technology. The first program is intended to review the issues related to applying the slurry approach as a complete cycle. The specific issues of applying the slurry approach to vehicles are being addressed in the second program. At the conclusion of these two programs, the merits and problems associated with storing hydrogen as a chemical hydride slurry will be known. As of the writing of this article, about two thirds of the work planned for each of these programs has been completed.

This article describes the chemical hydride slurry approach for production and storage of hydrogen and the major components required of the chemical hydride slurry cycle. We have included a discussion of the preliminary economic evaluation and an overview of the work performed for mobile applications of the technology.

CONCEPT

An essential feature of the chemical hydride slurry approach is the development of a relatively high-energy density hydrogen supply system based on the exothermic chemical reactions between metal hydrides and water. Hydrogen production via metal hydride and water reactions is a well-established industrial process. In fact, several groups of researchers have investigated the metal hydride/water reaction process to supply hydrogen for mobile power generators using fuel cells. This work has identified that reaction rate control, frequent on/off operation, and safety of the operation can be significant problems for high energy density operations.

One of the key technical challenges in the program is, therefore, to precisely control the metal hydride and water reaction. In our approach, the continuous organic slurry media will act as a path for dissipating heat that is generated from the hydride/water reaction. Furthermore, by controlling the surface chemistry of the organic media, the water/metal hydride reaction rate can easily be controlled.

Hydrogen storage in light metal hydride slurries has other benefits beyond reaction control. The hydride fuel can be handled as a liquid, simplifying transportation, storage, and delivery. Use of a slurry permits refueling similar to current gasoline filling stations, allowing the tank to be easily topped-off at any time. The hydroxide waste products produced by the hydrogen/water reaction can be removed from the onboard storage tank during the slurry filling operation. Both the hydride fuel and hydroxide waste product can be easily transported between the distribution centers and a central hydride slurry production plant.

Another essential feature of the chemical hydride slurry approach is that hydride, oil, and water used in the system are recycled. The way in which the metal hydride/water reaction would be used in a closed loop system for the storage and transmission of hydrogen is illustrated in Figure 1. The process consists of the following major steps:

(1) slurrying the metal hydride with a liquid carrier and transporting it to the point(s) of use,

(2) generating hydrogen on demand from the metal hydride/liquid carrier slurry at the point of use by adding water and then transporting the resulting metal hydroxide/liquid slurry back to the hydride recycle plant, and

(3) drying, separating, and recycling the metal hydroxide to the metal hydride at the centralized recycle plant and returning the liquid carrier for reuse.

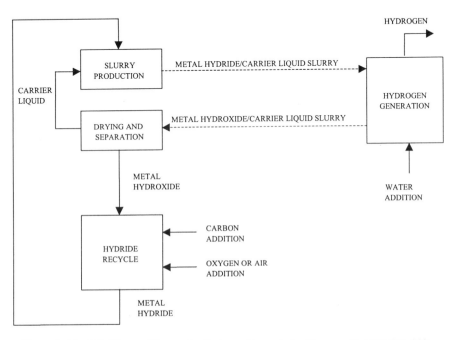

Figure 1. Simplified Process Diagram for Hydrogen Transmission/Storage with a Metal Hydride.

SELECTION OF THE LIGHT METAL HYDRIDE

A variety of metal hydrides react with water at ambient temperature to produce high purity hydrogen. Examples of reactions are:

$$LiH + H_2O \rightarrow LiOH + H_2$$
$$MgH_2 + 2H_2O \rightarrow Mg(OH)_2 + 2H_2$$
$$CaH_2 + 2H_2O \rightarrow Ca(OH)_2 + 2H_2$$
$$LiBH_4 + 4H_2O \rightarrow LiOH + H_3BO_3 + 4H_2$$
$$NaBH_4 + 4H_2O \rightarrow NaOH + H_3BO_3 + 4H_2$$

The hydrogen generation capability of these hydrides when reacted with water is outstanding. For example, the volume of H_2 (STP) produced by complete hydrolysis of

1 kg (2.2 lb.) of lithium hydride is 2800 liters (99 ft³), and by 1 kg (2.2 lb.) of lithium borohydride is 4100 liters (145 ft³).

In Table 1, the energy density of these hydrides when reacted with water is presented and compared to gasoline, as well as the storage of H_2 as a liquid, gas, and a reversible hydride. The energy densities of the reactive hydrides are given on the basis of the initial hydride mass. The energy densities of the hydride/water reaction are respectable when compared to gasoline or methanol, with $LiBH_4$ having the highest energy densities on both a mass and volume basis. The heat of reaction must be removed during the H_2 generation.

Table 1. Comparison of Metal Hydrides to Other Hydrogen Storage Methods, Gasoline and Methanol

Hydride	H₂ Volume Per Mass Hydride (STP L/kg)	Energy Density [5]		Water Reaction Enthalpy/HHV	Fraction Hydrolysis H₂ (kg H₂ per kg Hydride)	Hydride Density (gm/cm³)
		HHV/Mass, (MJ/kg)	HHV/Bulk Volume (MJ/L)			
Ca H₂[1]	1065	13.6	25.8	0.396	0.0958	1.90
Li H[1]	2820	36.0	27.7	0.388	0.254	0.77
Li B H4[1]	4116	52.5	34.7	0.212	0.370	0.66
Na B H4[1]	2370	30.2	32.5	0.157	0.213	1.074
Fe Ti H(1.6)[2]	170	2.2	11.9	0.122[4]	0.0153	5.5
Liquid Hydrogen[3]	—	142.0	9.9	—	—	0.07
Gaseous Hydrogen (34.5MPa, 300 K)	—	142.0	4.3	—	—	0.03058
Gasoline	—	47.9	36.2	—	—	—
Methanol	—	22.7	18.0	—	—	—

[1] Reaction with Water
[2] Dissociation by Heating
[3] Liquid Fuel
[4] Based on Dissociation Energy
[5] Assumes anhydrous hydroxide products

We have analyzed, both theoretically and experimentally, the reaction chemistry of a variety of metal hydrides and water, and the chemical stability of the organic carriers in contact with metal hydrides and spent hydrides. Since detailed hydrolysis reaction kinetics of the metal hydride/organic carrier slurry is not known, we conducted experiments using a high pressure (13.790 MPa or 2000 psi) and high temperature (232°C) vessel with temperature, pressure, and magnetic stirrer control capabilities (500 cm³ internal volume). Some of the selection criteria for the hydride follow.

- High specific energy density
- High hydrogen generation efficiency
- Relatively inert during storage before and after reaction with water
- Ease of regeneration
- Low costs

Table 2 summarizes the anticipated energy densities of the hydrides investigated. These energy densities are based on the assumptions that: the hydrides and product hydroxides will be contained in the same vessel; that the vessel will be 1.6mm thick; that the resulting hydroxide particle density will be defined with a 40% void fraction (40% of the volume of the particles will be empty); and that the container will be 20% larger than required to provide room for expansion or overfilling. The lithium hydride and the two boro-hydrides look the best from this analysis.

An additional concern that affects the choice of the hydride is that some of the hydroxides form hydrates, which tie up additional water in the water/hydride reaction. For instance, lithium hydride forms lithium hydroxide monohydrate if sufficient water is available. If the monohydrate is allowed to form, two moles of water will need to be supplied to the reaction for every mole of hydrogen produced by the reaction. Some of the other hydrides may require even more water to be supplied with the reaction. Sodium borohydride and lithium borohydride may form hydrates using several moles of water for each mole of hydrogen produced. This is undesirable for vehicular applications because the additional water must be carried rather than supplied from the fuel cell. In the case of the lithium hydride/water reaction the monohydrate does not form when the reaction occurs above 100°C (Kirk-Othmer 1995). The selection of the hydride and the design of the hydrogen/water reaction chamber are quite important when designing to minimize the system weight.

Table 2. Properties of the hydrides that were tested

Storage System		Wh/kg	%H_2	Wh/l	kg H_2/m^3
DOE Goal					
Liquid		3,355	10.1	929	28
Chemical Hydride	Slurry % Solids				
CaH$_2$	70	2,146	6.4	1,903	57
MgH$_2$	70	3,386	10.2	2,541	76
NaBH$_4$	60	3,978	11.9	2,223	67
LiH	60	4,641	13.9	1,919	58
LiBH$_4$	60	6,578	19.7	2,148	64

The following list represents the main conclusions of our evaluations.
- Borohydride regeneration is complex. Borohydrides form multi-hydrates.
- The sodium hydride reaction with water is pH limited and requires an additional reactant to force the reaction to completion.
- The magnesium hydride reaction rate with water was too slow. The reaction rate could be increased by addition of an acid or possibly by increasing the temperature of the reaction.
- Calcium hydride reacts well with water. However, it has a less desirable energy density of 1903 Wh/kg.
- Lithium hydride has a high energy density (4641 W-hr/kg). The hydride/water reaction is simple and can be performed to minimize the formation of hydrates.

Based on these findings, lithium hydride was selected as the primary hydride for our development program.

CHARACTERISTICS OF HYDRIDE SLURRY

The first issue in the preparation of the hydride slurry was the selection of the carrier fluid for the slurry. Light mineral oil has been selected for the carrier fluid because it is not chemically reactive with the metal hydride, it produces a relatively low vapor pressure, it is nontoxic, and it remains a liquid through the temperature range of –40 to 200 °C.

The next issue was to keep the hydride suspended within the slurry. The method we have chosen to accomplish this is shown in Figures 2 and 3. In Figure 2, a sketch is shown of two hydride particles, one surrounded by oil and one not. The oil layer inhibits the water access to the hydride and thereby controls the rate of reaction, which would otherwise be explosive. Dispersants are used to keep the particles in suspension. Figure 3 shows how the dispersant acts to hold the particle in suspension within the oil and further inhibit the reaction with water. The dispersant is made of an anchor group that attaches itself to the particles and a lyophile group that fends off adjoining particles and further protects the particle from unintentional reaction with water.

The amount of the dispersant and the size of the particles control the viscosity of the slurry. The selection of the viscosity will depend on the particular reactor design and the economics of the process. Tests have shown that the mineral oil can be recovered from the hydroxide slurry and reused. The mineral oil remains unchanged through the reaction.

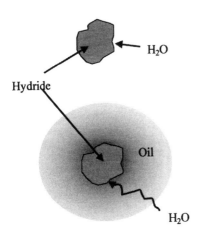

Figure 2. Hydride-Water Reaction Concept.

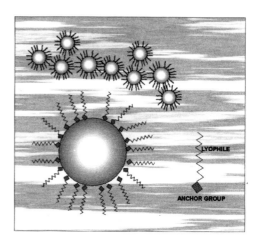

Figure 3. Hydrogen – Hydride Concept.

HYDROGEN GENERATION AT LOAD

Because of the reaction rate control afforded by the organic media, the hydrogen reactor can be a simple device. Water and hydride slurry are metered into the reactor, where they are thoroughly mixed to ensure complete reaction. This reaction goes to completion quickly, leaving a powdery waste. Hydrogen production rate is controlled by the injection rate of water and hydride. Heat released by the reaction is removed by the evaporation of some of the added water. No complicated control systems are needed to ensure proper and safe operation of the hydrogen reactor.

For a mobile application, the water required for thermal control and hydrogen reaction is provided by condensed water from the hydrogen fuel cell. Only a small reservoir of water is required for startup, makeup, and surge demand. Thus, the required water does not significantly affect the volumetric and gravimetric energy storage densities.

In the development of a hydride/water activation system, several design concepts were considered. These are:

- Single Tank Reactor
- Slurry Atomization Reactor
- Water Bathed Reactor
- Auger-aided Water Vaporizing Reactor

The single tank reactor, shown in Figure 4, is the simplest system. However, several problems exist for this system. The heat exchanger allows hot spots that increase the possibility of hydrocarbon contamination of the product hydrogen. It will also have a slow response to H_2 demand due to its large volume. Furthermore, it is likely that not all of the hydride will react, leaving a hazardous waste product. The large volume containing pressurized hydrogen is also a hazard.

The atomized slurry reactor, shown in Figure 5, was conceptualized to remove heat from 15 μm droplets by direct hydrogen heat transfer. This system is complicated and has a wear-prone slurry atomization system. The ½ m^3/sec H_2 flow rate needed for cooling is quite high. In addition, the heat exchanger may be fouled by dust. There is also the likelihood of poor hydride/water mixing reducing generation efficiency, and a large pressurized hydrogen volume is required.

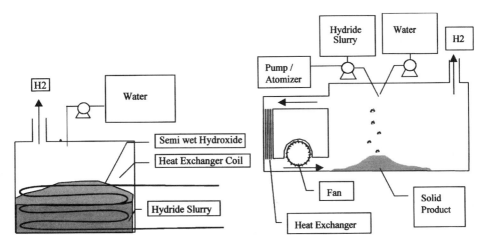

Figure 4. Simple One Tank Concept.   **Figure 5.** Atomized Slurry Reactor Concept.

In the water-bathed reactor configuration, shown in Figure 6, heat is removed by the recirculated flow of water. Excess water assures low reaction temperature and complete reaction in a relatively small reaction volume. In addition, cooling can be achieved by circulating a small amount of water to a water-to-air heat exchanger. Problems, such as the water soaked LiOH waste product and the weight of wasted water, push this concept outside the system goals of the vehicle program. The water could be separated by a filter or a filter press, but neither a filter or filter press system allows the concept to reach the weight goals. Also, unfiltered particles will wear the recirculation pump. This process may be the preferred configuration for a stationary hydrogen supply system however where additional water will not impose a weight penalty and where components can be chosen without regard to weight.

In the auger-aided reactor, shown in Figure 7, reactants are pumped to a mixing auger. At 300 rpm, the auger mixes, crushes particles, and eliminates foaming within the hydrogen generation reactor. The waste product contains 10% by mass of water and is a dry, free-flowing powder. About three times the stoichiometric water is added and vaporized by the heat of reaction to control the temperature. The water vapor content in the auger-aided reactor product hydrogen depends on the heat exchanger outlet temperature. Vapor condensation is slowed by the presence of hydrogen, increasing the size of the heat exchanger. The water vapor content could also be reduced by using the hydride as a desiccant. This hydrogen production system device achieves the weight and volume goals.

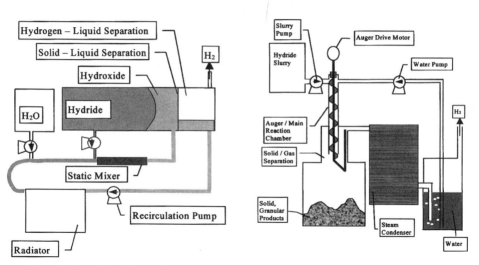

Figure 6. Water Bathed Reactor Concept.

Figure 7. Auger-Aided Reactor Concept.

Based on the preliminary analysis and testing of the various concepts discussed above, a prototype system to produce up to 3 kg/hr of hydrogen was designed. This system is shown schematically in Figure 8. A photograph of the test bed is shown in Figure 9. To produce the hydrogen, 0.5 l/min of a 60% LiH slurry flows into the auger reactor, along with 1.4 l/min of water for reaction and vapor cooling. The system produces up to 0.75 kg of hydrogen per run. A 1.6 gallon reservoir containing 60% LiH slurry, a 5.5 gallon water reservoir, and a 12 gallon hydroxide container make up the reactant and product volumes. The hydride and water pumps are computer-controlled. Data acquisition of pertinent pressures, temperatures, hydrogen flow, hydrocarbon, and water vapor content are recorded. The system is self-contained on a rolling cart.

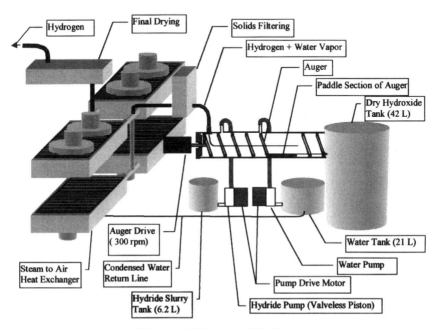

Figure 8. Schematic Of The Auger-Aided Reactor System.

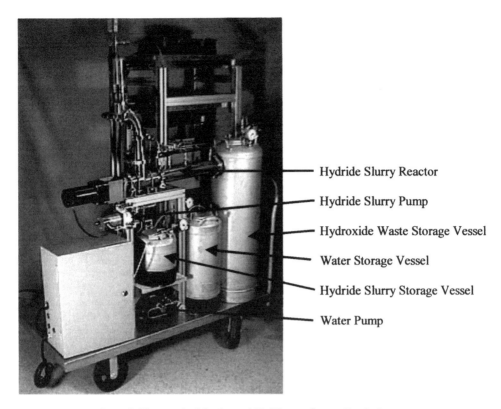

Hydride Slurry Reactor

Hydride Slurry Pump

Hydroxide Waste Storage Vessel

Water Storage Vessel

Hydride Slurry Storage Vessel

Water Pump

Figure 9. Photograph of the Auger-Aided Reactor System Test Bed.

A valveless ceramic piston pump is used for the LiH slurry and a gear pump is used for water flow. Three heat exchangers with 8-10" fans are used to condense the water from the hydrogen. Table 3 summarizes the energy, mass, and volumetric densities for the system, assuming storage of 15 kg H_2. To meet the design goals of 3355 Whr/kg and 929 Whr/l, our system must weigh less than 179.9 kg and must have a volume of less than 649.6 l. Table 3 shows that the system designed will meet the weight goal and exceed the volume goal.

HYDROXIDE WASTE RETURN

In stationary storage systems, the hydroxide waste will be stored as a slurry with excess water. It will be held in closed containers with active mixing and will be pumped into tank trucks or railroad cars and transported to a regeneration plant where it will serve as the raw material for the production of the hydride slurry. For vehicular storage the hydroxide will be carried as a dry powder that will be mixed with water to form a slurry as it is removed from the vehicular.

PRODUCTION OF HYDRIDE FROM HYDROXIDE

Electrochemical Reduction

Lithium hydride is formed by first reducing a lithium compound to lithium and then contacting liquid lithium metal with hydrogen above the melting point of lithium hydride

Table 3. System Mass and Volumetric Design Summary

		Wt (kg)	Vol (L)
65% Lithium Hydride Slurry		95.5	
Heat Exchanger	Direct Gas to Air HX Copper Tube, Aluminum Fin	31.38	82.5
Hydride Tank	Stainless	20	120
Hydroxide Tank	Plastic	12	310
Hydride Metering Pump	Valveless Piston	6.8	4
Water Metering Pump	Gear Pump + DC Control	3	2.5
Auger Drive Motor	1/8 Hp, 5.5:1 Bodine DC Gear Motor	4	2.3
Auger Construction	SS Materials	5	10
Total		177.7	531.3
System Goals (kg, l)		179.9	649.6
System Values (watt-hr/kg, watt-hr/liter)		3397	1136
Goals (watt-hr/kg, watt-hr/liter)		3355	929

(700°C). The reaction is highly exothermic and is controlled by controlling the rate of addition of hydrogen to the reaction vessel.

The current primary method for production of lithium is by electrolysis. Spodumene, the most plentiful lithium bearing ore, is beneficiated to 3 to 5% Li_2O and heated to 1000°C to convert it from its alpha form to its beta form. The beta form is treated with sulfuric acid to form Li_2SO_4. The Li_2SO_4 is water-soluble and is leached and reacted with sodium carbonate to form lithium carbonate. The lithium carbonate is then reacted with hydrochloric acid to form lithium chloride. Anhydrous lithium chloride is used to produce lithium metal by electrolysis (Austin 1984).

This process can be simplified if lithium hydroxide is used as a feed source. The lithium hydroxide can be contacted with hydrochloric acid to produce lithium chloride that can then be electrolyzed to create lithium metal.

We estimate that the electrochemical reduction of lithium chloride will require 0.34 kWh/gm-mole of lithium. This estimate is based upon the energy requirements for a similar electrolytic process for sodium metal production. Hydrogenation requires a further additional hydrogen stream that typically adds about $0.0016/gm-mole of lithium hydride production. We estimate that the cost of hydrogen, using the electrolytic route, will be $59/GJ of hydrogen for the electrical energy alone. For this reason, we have been exploring the carbo-thermal process for the reduction of lithium hydroxide to lithium hydride.

Analytical Evaluation of Carbo-thermal Reduction

The carbo-thermal reduction of lithium appears to offer a considerable cost benefit over the electrolytic production of lithium when performed for large quantities of lithium. We have not identified an existing commercial process for the carbo-thermal reduction of lithium oxide. However, the proposed process should be achievable since carbo-thermal reduction processes have been reduced to practice for magnesium. Since the bond energy of lithium oxide is less than the bond energy of magnesium oxide, the lithium reduction should actually be easier than that for magnesium. A carbo-thermal reduction plant for magnesium was built by Kaiser in Permanente, CA in the 1940's (Kirk-Othmer, 1995). It was eventually shut down because of difficulties in separating the product from its quenching medium. Since then several patents have been filed for improvements to this process step.

A preliminary design of the hydroxide to hydride regeneration system was analyzed to identify process stream conditions and to allow the major equipment components to be sized such that a capital equipment cost could be developed. One of the major benefits of this system is that hydrogen is produced as part of the process. This hydrogen is then contacted with the lithium metal later in the process to make lithium hydride. The system is shown in Figure 10. The analysis was conducted for both lithium hydroxide and calcium hydroxide regeneration.

The material and energy balances for the two metals were determined for a plant supplying hydrogen to 250,000 cars. Such a plant would produce enough slurry to supply 13 tons of H_2/hr. It would be small relative to typical chemical engineering projects, however. The first Fluid Catalytic Cracking (FCC) plant was three times larger than this proposed system and today's FCC plants are 25 times larger.

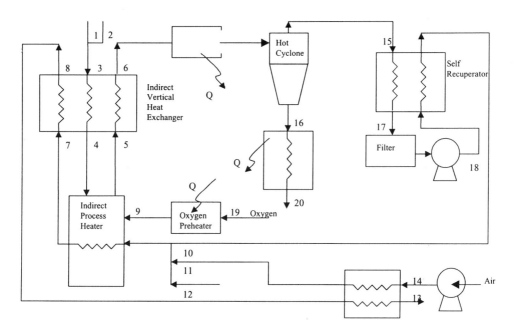

Figure 10. Hydroxide Regeneration System Flow Schematic.

Lithium hydroxide, stream 1, is combined with carbon for the reduction and fuel, stream 2, to form stream 3, and is fed to the top of an indirect vertical heat exchanger, which preheats the incoming reactants while cooling the stream containing the lithium hydride, streams 5 and 6. The possibility for removing heat from the indirect-fired process heater is also provided with streams 7 and 8. The hot preheated and partially reacted reactants, stream 4, enter the reduction reactor in which they are heated indirectly to the reaction temperature by combustion of the recycled carbon monoxide, stream 10, and additional fuel, stream 12, with preheated air, stream 11. The possibility of adding direct heat to the reactor is accomplished by adding oxygen to the reduction reactor by stream 9. The products of reduction leave the reduction reactor through stream 5. Within the reactant preheater, the lithium hydride is formed through the non-equilibrium kinetics as the mixture of lithium, hydrogen, and carbon monoxide is cooled. Additional heat is taken out of the product stream for the generation of electrical energy, which is added back into the reduction reactor to reduce the additional fuel.

The product, lithium hydride, is separated from the carbon monoxide in the hot cyclone, stream 16. This is further cooled to produce additional power, which is also added to the reduction reactor. The hot carbon monoxide, stream 15, is passed through a self recuperator to get a cold stream of CO, which could have a barrier filter installed to remove all the lithium hydride and a blower to circulate the CO, stream 18. This stream is reheated with the incoming CO and fed into the indirect process heater as discussed above. The hot combustion products leaving the solids preheater, stream 8, are used to preheat the combustion air and produce power, which is fed back into the reduction reactor. The energy efficiency of the hydrogen storage is obtained by dividing the heat of combustion of the hydrogen in the metal hydride by the heat of combustion of the carbon used for the reduction and the additional fuel. The results are lithium (52.1%) and calcium (22.9%).

During the past few months, we have been evaluating additional processes for carbo-thermal reduction of lithium hydroxide. We have been using the ASPEN Plus™ code (Version 10) developed by Aspen Technologies, Inc. The ASPEN Plus™ code provides a convenient means for modeling process designs. The results of our analyses indicate that lithium hydroxide may be reduced with carbon at temperatures less than 1500°K if the products of the reaction are swept out of the reactor as they are produced. We are currently using a high temperature furnace to evaluate the process steps. The furnace results confirm the predictions and provide a potentially much easier method for reducing lithium than we had evaluated in our earlier analysis where the peak temperature was 1850°K.

Experimental Evaluation of Carbo-thermal Reduction

Experiments are underway to evaluate the critical steps required in the carbo-thermal production process. In this process the lithium hydroxide is reacted with carbon at an elevated temperature. This process could yield a number of products depending upon the relative kinetics and thermodynamics. The potential pathways are shown in Figure 11.

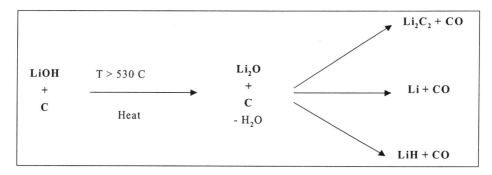

Figure 11. Potential Pathways.

A small furnace has been built for evaluating the regeneration process. A graphite crucible is used to contain the sample. A molybdenum heating element allows us to heat the crucible to the desired temperatures. The furnace is mounted in a quartz tube, as shown in Figure 12. The top and the bottom of the enclosure is made from aluminum plate. Thermocouples are provided in the crucible and in the chamber. A sight glass is mounted on the top cover to give a direct view of the hot section of the crucible. Argon cover gas is piped in and the off gases are filtered and continuously analyzed for oxygen, carbon dioxide and carbon monoxide. A detail of the heater and crucible arrangement is shown in Figure 13. Figure 14 is a photograph of the completed furnace in operation.

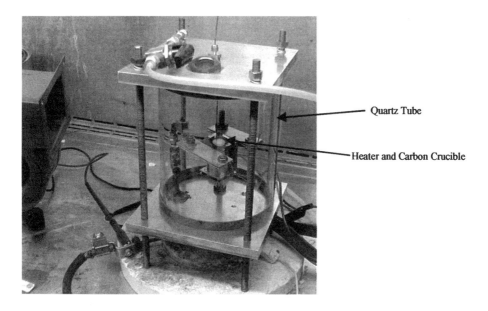

Quartz Tube

Heater and Carbon Crucible

Figure 12. Furnace Arrangement.

Figure 13. Heater and Crucible Detail.

Figure 14. Furnace in operation.

Table 4. Reaction Products

	FURNACE	FILTER	GAS PRODUCTS
	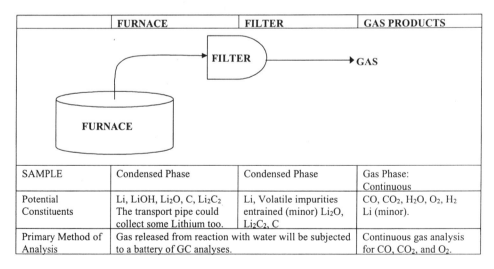		
SAMPLE	Condensed Phase	Condensed Phase	Gas Phase: Continuous
Potential Constituents	Li, LiOH, Li_2O, C, Li_2C_2 The transport pipe could collect some Lithium too.	Li, Volatile impurities entrained (minor) Li_2O, Li_2C_2, C	CO, CO_2, H_2O, O_2, H_2 Li (minor).
Primary Method of Analysis	Gas released from reaction with water will be subjected to a battery of GC analyses.		Continuous gas analysis for CO, CO_2, and O_2.

Li, LiH and Li_2C_2 are the potential final products. The possible sample/product distribution and the analysis methods are shown in Table 4.

The condensed-phase samples will be reacted with water to determine the abundance of various lithium compounds. If lithium carbide is present, acetylene will be produced when the lithium carbide is reacted with water. Other compounds (LiH and Li) should release hydrogen when contacted with water. These gases will be measured with a gas chromatograph.

The initial tests performed with this furnace have been inconclusive. We are currently performing shakedown and quality control tests to identify the capabilities and limitations of the furnace.

ECONOMICS

The preliminary economics for the process have been obtained by first developing a capital cost for the process equipment and then estimating the operating cost to define the needed sales price of the metal hydride for the required after tax return on the investment.

The capital equipment costs for the process are shown in Table 5 for the lithium process. These estimates, as well as the operating cost estimates, were obtained using standard chemical engineering practice (Ulrich, 1984). The operating cost assumptions are shown in Table 6.

The sensitivity of the cost of the hydride and the rate of return as a function of plant size and carbon cost is shown in Figures 15 and 16 for lithium. In Figure 15, the cost of hydrogen is plotted versus the plant size for four values of the cost of carbon. For a 250,000 car-per-day plant, the cost of hydrogen is on the order of $3.42/GJ ($3.61/MMBtu) at a carbon cost of $0.022/kg ($0.01/lb) and a fixed after tax return on the investment of 15 percent. In Figure 16, the effect of plant size and carbon cost for a fixed hydrogen cost on the rate of return is shown. In this case, if the hydrogen can be sold for a value of $4.33/GJ ($4.57/MMBtu), the return to the investors can range from 15 to 65 percent, depending on plant size and carbon price. The same trends are seen for calcium.

Our program plan is to reevaluate the economics of the system at the end of the program. At that time, we will have improved process designs for each of the process steps in the regeneration, slurry production, hydrogen generation, and transportation.

Table 5. Capital Cost - Lithium Hydride Regeneration for a 250,000 Car Per Day Plant

	Total cost ($)
Furnace Cost, base 70m3	9,236,116
Solids preheater, 70 m3	9,236,116
Condensor, base 100MW	-
Hydride Reactor, Base 35m3	720,417
Blower, H2 from sep.base, 75m3/s	270,254
Steam Turbine Generator	25,693,663
Cent Slurry sep.	189,413
Hydride cooler, base 70 m3	9,236,116
Heat Exch/recuperator, base 20e9J/s	2,814,328
Hydrocarbon Decomp, base 100MW	-
Sum, Total Cost	57,396,424

Table 6. Operating Cost Assumptions for a 250,000 Car Per Day Plant

Carbon	Variable, $0.64 to 1.58/GJ
Fuel	$2.37/GJ

Labor
  -Operators
  -Supervision & Clerical
	
	25 at $35,000/yr
	15% of Operators

Maintenance & Repairs	5% of Capital
Overhead	50% of Total Labor and Maintenance
Local Tax	2% of Capital
Insurance	1% of Capital
G&A	25% of Overhead
Federal and State Tax	38% of Net Profit

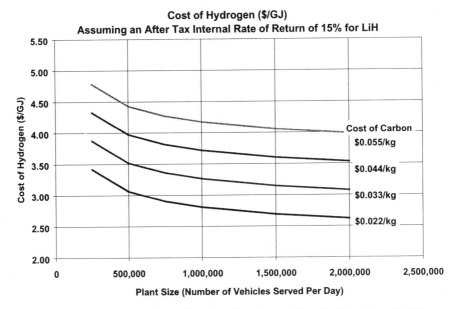

Figure 15. Sensitivity Of Hydrogen Cost To Carbon Cost And Plant Size For Lithium Hydride.

CONCLUSIONS AND REMAINING ISSUES

The chemical hydride slurry approach continues to offer many advantages over other competing methods of storing hydrogen. The storage medium is stable at normal temperatures and pressures. If hydrogen is needed at an elevated pressure, the reaction vessel can be constructed to provide supply hydrogen at the elevated pressure. The slurry and its storage vessel exceed the gravimetric and volumetric energy density goals set by DOE for liquid storage. When the control system and pumps are added for a vehicular application, the system appears to be capable of meeting the gravimetric energy density goal and exceeding the volumetric energy density goal. Future design improvements could allow the system to exceed the design goals.

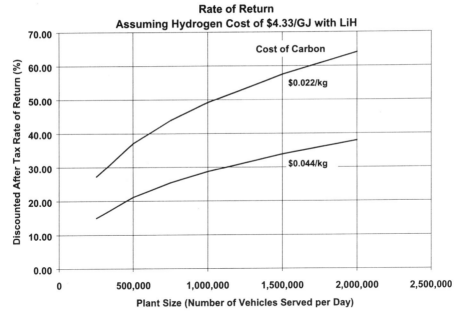

Figure 16. Sensitivity Of Rate Of Return To Carbon Cost And Plant Size For Lithium Hydride.

ACKNOWLEDGMENTS

This work is supported by the U.S. Department of Energy (DOE) (Award Numbers DE-FC36-97GO10134 and DE-FC02-97EE50483), the California South Coast Air Quality Management District (AQMD Contract No. 98111), the Southern Illinois University, and the Thermo Power Corporation. The opinions, findings, conclusions, and recommendations are those of the author(s) and do not necessarily represent the views of DOE or AQMD. DOE, AQMD, their officers, employees, contractors, and subcontractors make no warranties, expressed or implied, and assume no legal liability for the information in this report. DOE and AQMD have not approved or disapproved this report, nor have they passed upon the accuracy or adequacy of the information contained herein.

REFERENCES

Austin, George T., *Shreve's Chemical Process Industries* , Fifth Edition, McGraw-Hill Book Company, New York, 1984

Block, D. and I. Melody, "Plans For A U.S. Renewable Hydrogen Program," Florida Solar Energy Center, *Proceedings of 10th World Hydrogen Energy Conference* , Cocoa Beach, FL, (June 20-24, 1994).

Breault, Ronald W., Rolfe, J., and McClaine, A., "Hydrogen For The Hydrogen Economy," *24th Coal Utilization and Fuel Systems Conference, Clearwater* , FL,(March 8-11, 1999).

Breault, Ronald, "Advanced Chemical Hydride Hydrogen Generation/Storage System for PEM Fuel Cell Vehicles," to be published in the *DOE Office of Advanced Automotive Technology FY 1999 Annual Progress Report*.

Gordon, Sanford and Bonnie J. McBride, *Computer Program for Calculation of Complex Chemical Equilibrium Compositions and Applications* , NASA Reference Publication 1311, 1994

Kirk-Othmer, *Encyclopedia of Chemical Technology, Fourth Edition* , Volume 15, p450, 638, John Wiley & Sons, New York, 1995

Padró, C.E.G., and V. Putsche, *Survey of the Economics of Hydrogen Technologies* , National Renewable Energy Laboratory, NREL/TP-570-27079, (September 1999).

Ulrich, *A Guide To Chemical Engineering Process Design And Economics* , John Wiley & Sons, (1984).

ADVANCES IN LOW COST HYDROGEN SENSOR TECHNOLOGY

Rodney D. Smith, II
DCH Technology, Inc.
Valencia, California

David K. Benson
National Renewable Energy Laboratory
Golden, Colorado

J. Roland Pitts
National Renewable Energy Laboratory
Golden, Colorado

Barbara S. Hoffheins
Oak Ridge National Laboratory
Oak Ridge, Tennessee

INTRODUCTION

Hydrogen may be emerging as the fuel of choice for an energy carrier. It can be stored, handled, reacted or combusted to deliver large quantities of energy to an end use safely, conveniently, and efficiently with very little environmental impact. However, it is a combustible gas, and the public has been sensitized to dangers associated with its use. Safe practices and codes for handling hydrogen will require convenient and reliable methods of detecting hydrogen leaks in spaces where combustible or explosive concentrations may be reached. The U. S. Department of Energy has undertaken many of the long-range tasks associated with bringing a new energy carrier into widespread use and has initiated study of new sensor technology that will meet the requirements imposed by new technology. DCH Technology, Inc., has taken the lead in the private sector to develop and make available to the public new hydrogen sensors that will meet the code requirements and allow the advent of hydrogen based fuels.

Expanded use of hydrogen in the public domain brings new requirements for safety monitoring, which have not been considered until recently. For instance, the use of hydrogen for a transportation fuel will necessitate the outfitting of each vehicle and each fueling area with multiple sensors to detect low concentrations of hydrogen and to initiate a set of hierarchical actions such as setting off alarms, activating fans, etc. prior to the onset of the explosive limit. The sensors must be rugged, reliable, and inexpensive enough to incorporate several into each vehicle. Additionally, the sensors need to be lightweight and have minimal energy requirements themselves. In order to meet such challenges, DCH Technology, in conjunction with the National Renewable Energy Laboratory (NREL) and Oak Ridge National Laboratory (ORNL), is developing and commercializing solid-state hydrogen sensors. They are designed to be inexpensive, small, and chemically inert. The technologies are based upon either chemochromic or resistance changes in the properties of thin films in the presence of hydrogen. The NREL Fiber Optic (chemochromic) sensor requires no electrical power at the sensing point and is ideal for high electromagnetic environments. Furthermore, a modification of the fiber optic sensor has shown promise as an analytical tool for measurement of diffusible hydrogen in welded steel. The ORNL thick film (resistive) sensor is versatile and can operate from a small battery. Data from combinations of multiple sensors can be fed into a central processing unit via fiber optics or telemetry to provide hydrogen situational awareness for small and large areas. These sensor technologies and their current state of development are presented below.

FIBER OPTIC HYDROGEN SENSOR (FOHS)

The ability to detect hydrogen gas leaks economically and with inherent safety is an important technology that could facilitate commercial acceptance of hydrogen fuel in various applications. In particular, hydrogen fueled passenger vehicles will require leak detectors to signal the action of various safety devices. Such detectors will be required in various locations within a vehicle, wherever a leak could pose a safety hazard. It is therefore important that the detectors be very economical. For purposes of early detection a fast response time (≤ 1 second) is also desired. An optical fiber coated with a thin film of a chemochromic (color change induced by a chemical reaction) material offers the possibility of meeting these objectives.

Chemochromic materials such as tungsten oxide and certain lanthanide hydrides can react reversibly with hydrogen in air while showing significant changes in their optical properties. Thin films of these materials applied to the end of an optical fiber have been used as sensors to detect low concentrations of hydrogen in air. The coatings include a thin layer of gold in which a surface plasmon (Raether 1988) is generated, a thin film of the chemochromic material and a catalytic layer of palladium that facilitates the reaction with hydrogen. The gold thickness is chosen to produce a guided surface plasmon wave between the gold and the chemochromic material.

A dichroic beam splitter separates the reflected spectrum into a portion near the resonance and a portion away from the resonance and directs the portions to two separate photodiodes. The electronic ratio of these two signals cancels most of the fiber transmission noise and provides a stable hydrogen signal.

A fiber optic sensor based on the palladium catalyzed reaction of amorphous tungsten oxide and hydrogen was first proposed by Ito (Ito and Kubo 1984). This simple sensor design was found to be adequate in terms of sensitivity but too slow in response time for the intended use. A different design using a surface plasmon resonance (SPR) configuration was therefore investigated (Benson et al. 1998). The SPR shifts in response to subtle changes in the refractive index of the coating. This shift can be monitored to give a faster response.

FOHS: Experimental

The thin film sensors were deposited on 20 mm right angle prisms for characterization. The test apparatus was sealed to protect the film from the surrounding air and to expose the films to predetermined gas mixtures, as shown in Figure 1. The sensors consisted of three layers: 40 nm gold, 600 nm WO_3 and 3 nm palladium. Figure 2 shows a set of spectra from a time series taken during exposure to 5% hydrogen in air at room temperature. The inset shows the change in reflected intensity at the resonant wavelength over time. The change in signal amplitude is approximately exponential with a time constant of about 20 seconds.

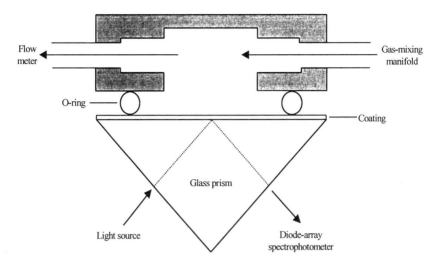

Figure 1. Schematic cross section through the SPR sensor sample holder.

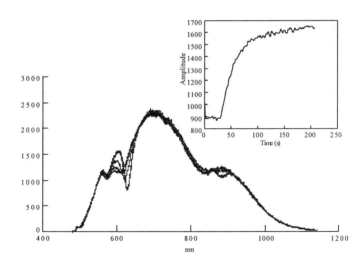

Figure 2. Selected reflectance spectra from a time series during exposure to 5% H_2 in air. The inset shows reflected intensity at 627 nm versus time. The coating is 40 nm Au/600 nm WO_3/3 nm Pd.

Tungsten oxide readily exchanges water vapor with the atmosphere. Water absorbed onto the nanoporous structure of the WO_3 increases its refractive index. Conversely, exposure of the sensor to dry air causes loss of absorbed water from the film that mimics exposure to hydrogen. Attempts were made to retard water absorption by applying a protective layer of polytetrafluoroethylene over the palladium. This modification retarded the exchange of water vapor without decreasing the sensitivity to hydrogen. However, the sensor was still susceptible to long term drift under conditions of changing humidity.

Increasing the film thickness of the palladium layer was also unsuccessful. Figure 3 shows the response of such a sensor (17 nm Ag/330 nm WO_3/100 nm Pd) during exposure to 0.9% hydrogen in air. The response time constant is a few seconds. Over time, however, this response time increased dramatically. Figure 4 shows the measured response time for the sensor over a period of 2 days. The time constant is seen to increase from a few seconds to more than 200 s, approximately increasing in proportion to the square root of time the sensor was exposed to the test gases.

More recent experiments have indicated that the observed degradation may be due in part to poisoning of the palladium catalyst by airborne contaminants such as CO or H_2S. Figure 5 compares the changes in response over time for simple non-SPR mode sensors with and without an additional protective coating of WO_3. The top curve shows the time constants of a sensor with an unprotected Pd film after exposure to ambient air. As seen in the figure the time constant increases dramatically within a few days. The curve on the bottom represents a sensor with the additional protective coating. Although the data are preliminary, they suggest that the extra WO_3 coating is protecting the Pd layer from being poisoned. This effect is caused by the oxidative catalytic properties of the tungsten oxide.

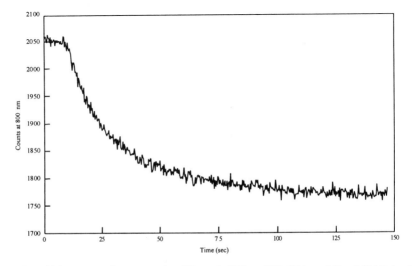

Figure 3. Initial response of sensor coating (17 nm Ag/330 nm WO_3/100 nm Pd) to 0.9% H_2 in air.

Alternative materials for the sensor films have also been investigated. Lanthanide and other rare earth hydrides have been shown to undergo reversible transitions between di- and tri-hydride upon exposure to hydrogen in air (Griessen et al. 1997). Thin films of these materials exhibit dramatic changes in optical properties that may make them suitable for hydrogen sensing. Figure 6 shows reflectance measurements of a yttrium hydride film upon repeated exposure to 0.45% hydrogen in air. The time constant for this film was approximately 6 seconds. A YH_2 film incorporating the SPR design may show greater sensitivity and speed. Further work is underway with these hydride sensor films.

FOHS: Conclusions

The thin film fiber optic reflective sensor based on palladium catalyzed WO_3 appears to work well for detecting hydrogen, but its response time is too slow for critical applications. The sensitivity and speed of hydrogen detection can be enhanced by using the sensor film in a SPR configuration. However, in such a configuration the rapid desorption of water vapor can lead to a false indication of hydrogen whenever the sensor is exposed to a dry gas. Increasing the thickness of the palladium layer is ineffective in preventing degradation of the sensor. The use of protective coatings has shown promise for retarding this degradation, which may be due to poisoning of the catalyst by airborne contaminants.

Preliminary experiments with yttrium hydride showed increased speed and sensitivity in a simple sensor configuration. Optimization of this design, including the use of SPR, may provide the needed combination necessary for the intended safety application. Research is continuing to determine the ruggedness and durability of the hydride films.

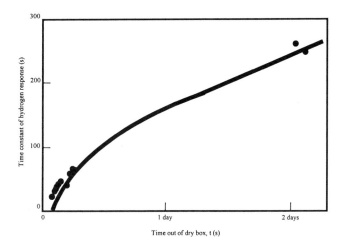

Figure 4. Sensor response time constant for film (17 nm Ag/330 nm WO_3/100 nm Pd) over a period of 2 days. The sample was exposed to 5% H_2 in air for periods of about 2 minutes, then to dry air. The response time constant was measured away from the SPR, at 800 nm. The fitted curve varies as the square root of time.

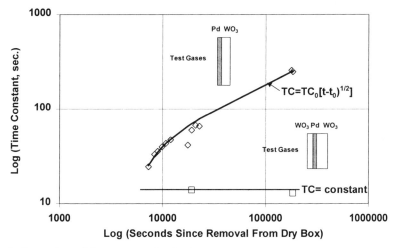

Figure 5. Comparison of time constants for simple non-SPR sensor films with and without additional protective coating of WO_3.

153

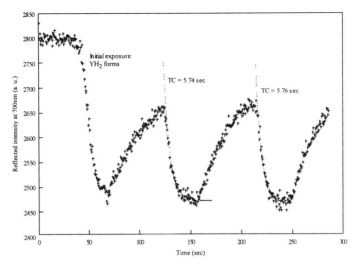

Figure 6. Initial response measurements for a yttrium-hydride film (17 nm Ag/100 nm Y/20 nm Pd) to 0.45% H_2 in air. The first reduction in the signal corresponds to conversion of Y metal to YH_2, and subsequent cycles to increases in hydrogen content toward the tri-hydride, YH_3. The response was measured in reflection (no SPR) at 700 nm. The time constant of the hydrogen response is about 5.7 seconds

FIBER OPTIC WELD SENSOR (FOWS)

This work was performed in corroboration with the Center for Welding, Joining and Coatings Research at Colorado School of Mines, Golden, CO. The concept of the fiber optic hydrogen sensor can be applied to the measurement of diffusible hydrogen content in welded steel. The rate of hydrogen diffusion from the deposited metal of a welded sample is proportional to the initial concentration, which can then be calculated. Diffusion rates for a typical sample can be determined in less than one hour. The sensors are also relatively inexpensive and simple to operate. These factors make the use of the sensors for diffusible hydrogen measurements attractive when compared to present testing procedures.

FOWS: Experimental

The sensors for these experiments consisted of polymer optical fibers coated with thin films of WO_3 and palladium; the films were deposited by vacuum evaporation. The average thickness of the WO_3 films was 500 nm while the thickness of the palladium films varied from 3 to 30 nm. The change in optical properties was measured using an infrared (IR) reflectometer equipped with a laser photodiode emitting at 850 nm. The reflectance was measured in units of reflected power (μW). The typical response of an unreacted sensor was approximately 1.5 μW. The sensors were extremely sensitive to hydrogen and would fully react to a mixture of 1% hydrogen in argon in only a few minutes, resulting in a power decrease of approximately 1.3 μW.

The response of the sensor to hydrogen diffusing from a weld was investigated by designing a sensor housing for attachment to the base plate of a welded sample as shown in Figure 7. The sensor was placed directly adjacent to the weld deposit. Using this design it was found that the amounts of hydrogen diffusing from the base plate were too small for a detectable response. The design was therefore modified to sample from the curved surface of the weld deposit using a soft rubber gasket adapter. The gasket conformed to the curvature of the weld bead, as shown in Figure 8.

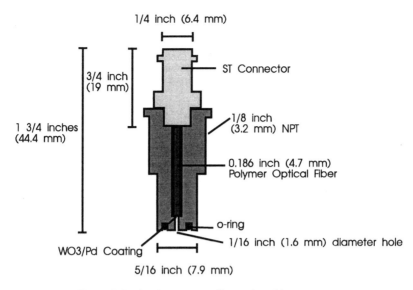

Figure 7. Design for prototype fiber optic weld sensor.

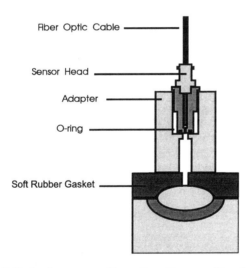

Figure 8. Design for prototype fiber optic weld sensor with soft rubber gasket.

The experiments were conducted using ASTM A36 steel samples. The samples were gas metal arc welded (GMAW) with 0.1% H_2/argon, quenched in ice water and stored in liquid nitrogen (LN_2) until analysis. Data was collected using a General Purpose Instrumentation Bus (GPIB) interface connected to a laptop computer. Measurements were made every minute. The results shown in Figure 9 indicate that the amount of hydrogen diffusing from the weld deposit was more than adequate for detection. As shown by the response curves the hydrogen from the weld deposit was still detectable by the sensor after a period of five hours.

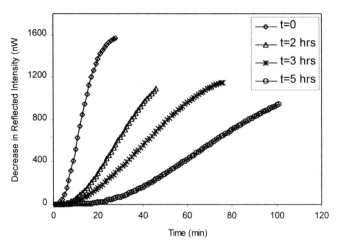

Figure 9. Hydrogen response data as a function of time after quenching for gas metal arc welded ASTM A36 steel (0.1% H_2/Ar shielding gas) using prototype fiber optic weld sensor.

The next set of experiments was designed to gain quantitative measurements from the steel samples. For these experiments HSLA 100 samples were used and gas metal arc welded (GMAW) using different levels of hydrogen in the shielding gas. The samples were welded in duplicate; one set of samples was analyzed by the standard gas chromatography (GC) method to generate results in ml/100 g weld metal. The other sample set was analyzed using the sensor.

An attempt was made to correlate the slope of the sensor response curve to the initial diffusible hydrogen concentration in the sample. The steady state portion of the curve could be assumed to be proportional to the flux of hydrogen from the weld metal. To investigate this possibility theoretical curves were generated using an equation derived from the error function erf(x).

The diffusion was assumed to occur from a semi-infinite plane sheet with a uniform initial concentration (C_o) throughout and a constant surface concentration (C_1) of zero. For this case the solution to the error function equation takes the form:

$$ M_t = 2 C_o \left(\frac{Dt}{\pi} \right)^{1/2} $$

where M_t is the amount of diffusing substance, D is the diffusion coefficient for the substance in a particular medium, C_o is the initial concentration and t is time (Crank, 1992). Solutions were generated for different initial concentrations as a function of time; the results were adjusted for the average amount of weld metal per sample and multiplied by the ratio of surface area sampled to total area.

The initial data gathered with the gasket adapter yielded a correlation between the actual samples and the theoretical values that was unsatisfactory. It was assumed that the surface area actually sampled was greater than estimated due to a poor gasket seal. A new sensor adapter was therefore designed as shown in Figure 10. The original sensor head was again used; the adapter was threaded at both ends. One end accepted the sensor head and the other a proboscis-like tip. The tip could be inserted into a hole drilled in the weld metal.

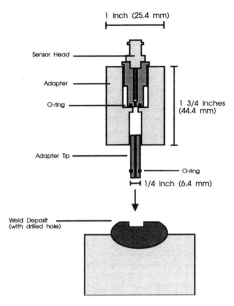

Figure 10. Proboscis adapter design for prototype fiber optic weld sensor.

The experiments were repeated with the new adapter. Three sets of samples were welded and analyzed with three different fiber sections. Each section was calibrated prior to the analyses. The slopes for each response curve were calculated and compared to the theoretical data. A best fit was obtained with an apparent diffusion coefficient of D_{eff} = 7.5 x 10^{-5} cm^2/sec. The results are presented in Figures 11-13.

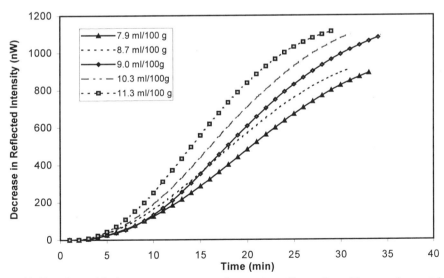

Figure 11. Experimental hydrogen response curves for prototype fiber optic weld sensor (gas metal arc welded HSLA 100 steel)

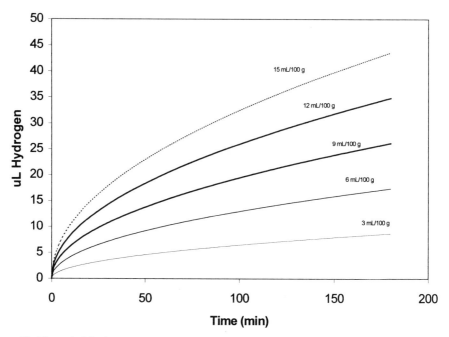

Figure 12. Theoretical hydrogen evolution curves using diffusion equation for semi-infinite plane sheet (based on erf(x); five different initial diffusible hydrogen concentrations; D_{eff} = 7.5 x 10^{-5} cm^2/sec).

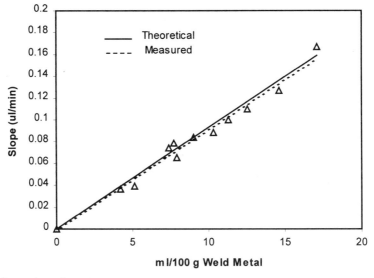

Figure 13. Comparison of experimental and theoretical data for fiber optic weld sensor (gas metal arc welded HSLA 100 steel; Deff = 7.5 x 10^{-5} cm^2/sec; r = 0.989).

FOWS: Conclusions

The fit to the theoretical data is quite good, especially when the non-uniformity of the weld deposits is taken into account. These results indicate that the sensor could be used as an analytical tool for diffusible hydrogen measurements in welded steel. The advantages of using the sensor over performing the standard test are numerous. The sensor response is extremely rapid, with results obtained within ½ hour. The test may be performed on the actual structure in the field, as opposed to a laboratory sample. The sensor is extremely sensitive and can detect hydrogen down to levels of 1 or 2 mL/100 g, as required for higher strength welded steels. The equipment used is relatively inexpensive, and can be designed for use by welding personnel. Further research is needed to provide complete non-destructive testing of the weld by insuring an adequate seal without drilling.

It is anticipated that refinement of the sensor design will provide the welding industry with a powerful tool for control of hydrogen damage in welded structures, especially those structures manufactured from high strength alloys where such control is most critical.

THICK FILM HYDROGEN SENSOR (TFHS)

Researchers at Oak Ridge National Laboratory (ORNL) have developed a solid state hydrogen sensor using thick film technology (Lauf et al., 1994, Hoffheins et al. 1995). The sensing mechanism of the ORNL sensor relies on the reversible absorption of atomic hydrogen into palladium metal. Changes in hydrogen concentration in the palladium matrix lead to corresponding changes in the electrical resistance of the palladium that are easily measured.

The design consists of four palladium resistors (or legs) that are arranged in a Wheatstone bridge configuration, as shown in Figure 14. The circuit is based upon three thick film components. Each layer is separately printed and fired on a ceramic substrate. The layers include a conductor that joins the palladium segments and provides connection points for power and signal circuitry, a palladium resistor that serves as the sensing layer and consists of four serpentine segments, and a passivator that forms a hydrogen impermeable barrier over two of the palladium resistors. The two passivated legs serve as reference resistors and thus compensate for changes due to temperature variation. The palladium resistor material used in fabricating the sensors was developed and patented by DuPont Electronics (Felten 1994).

TFHS: Experimental

The TFHS has been tested under a wide variety of conditions (Hoffheins et al. 1997, Hoffheins et al. 1998). It has shown good response to a range of hydrogen concentrations at temperatures between 0 and 200°C, and in dry and humid environments. Figure 15 shows the response of a TFHS to increasing concentrations of hydrogen in air from 0.2 to 2%. The sensor was placed in a small test chamber (50 cm^3). Hydrogen was added to air in 0.1% increments, while flow was maintained at a constant rate. Each step was 60 seconds. The sensor is insensitive to the presence of hydrogen below the 0.2% level. From 0.2% up to 2% hydrogen the response is linear. For each successive increment of hydrogen, sensor output increased and leveled off in 7 seconds. The time constant of the test chamber is estimated at about 4 seconds. Thus the actual sensor time constant could be as little as three seconds to reach maximum output for the indicated increases in hydrogen concentration.

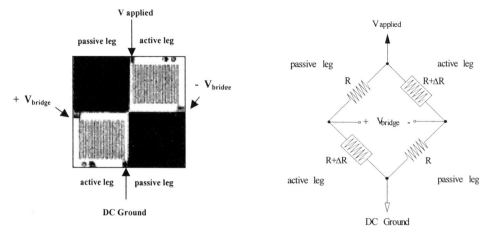

Figure 14. Thick film hydrogen sensor chip and schematic representation.

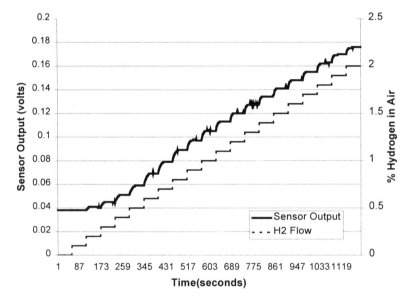

Figure 15. Thick film hydrogen sensor response to increasing concentrations of hydrogen in air.

The operating environment of the sensor can lead to exposure to a multitude of gases and vapors. Test results showing a decrease in sensor response over time may be attributable to adhesion of water vapor or other species that form a temporary barrier to hydrogen gas. Hydrogen permeable protective coatings for the palladium layer are a possible solution to improve response time reliability. Candidate materials include SiO_2, Teflon[TM], thick film dielectrics, acrylic and silicone rubber. Researchers at ORNL are evaluating several of these materials, as well as testing the sensor against gases such as CO_2, CH_4, NH_3, O_2 variations, water vapor, acetylene, and chemical vapors to which the sensor could be exposed. Additional considerations under evaluation are susceptibility of the sensor and associated wiring and electronics to electromagnetic interference. Also, sensor ruggedness and durability are being characterized.

TFHS: Applications

DCH Technology of Valencia, California is continuing its economic evaluation for the purpose of commercializing the TFHS. This analysis has so far identified three initial target markets in particular and several others in general. The automotive field of use is the first market of interest. There is a demand in the automotive industry for a rugged, low cost device for use in vehicles powered by fuel cell and hydrogen fueled combustion engines. Research is continuing to meet the design specifications for automotive applications using the thick film hydrogen sensor, such as fast response time and durability. Several automobile manufacturers have expressed interest in testing the sensor for their hydrogen vehicles.

The concept of a hydrogen safety badge is the second market area. The thick film sensor could be coupled with a small circuit board and battery for use as a safety badge in areas where a hydrogen hazard might exist. If hydrogen is detected in amounts over a predetermined alarm threshold, both audio and visual alarms are triggered on the badge. DCH Technology has designed and produced a prototype board that was displayed at the National Hydrogen Association Meeting April 7-9, 1999. The entire badge is about the size of a standard business card.

The third application is the coupling of the sensor with a radio tag for wireless monitoring. This will be useful in such areas as refineries and utilities where the potential for hydrogen leaks exists, but existing technology requires thousands of feet of cable to be installed. The dimensions and profile of the sensor and associated electronics would allow for remote monitoring and transmission of signals over a long distance.

Other applications include industrial safety monitoring, primarily in semiconductor plants, metals processing and hydrogen generation plants. A series of agreements for beta testing are in negotiation for these markets.

TFHS: Conclusions

Research is continuing to develop and refine the sensor design. The TFHS is inherently rugged and inexpensive, making it attractive as a wearable personnel safety device or as a monitoring device for hydrogen fueled vehicles. Further development of the sensor will address the issues and requirements for automotive safety and other applications.

SUMMARY

DCH Technology is commercializing state of the art hydrogen sensors developed by Oak Ridge National Laboratory and the National Renewable Energy Laboratory. This blend takes the scientific knowledge resident in these laboratories and creates applicable, timely, and meaningful commercial products that assist the development of the hydrogen fuel economy. These sensors could help carry this development into the public domain.

The authors wish to acknowledge the support of the United States Department of Energy under contracts #DE-AC36-98-G010357 (National Renewable Energy Laboratory) and #DE-AC05-96OR22464 (Oak Ridge National Laboratory).

REFERENCES

Raether, H. 1988, *Surface Plasmons.* Springer Verlag, Berlin.

Ito, K. and Kubo, T. 1984, Gas detection by hydrochromism, *Proc. Mat. Sci. Symp.* 403:153.

Benson, D. K., Tracy, C. E. and Bechinger, D. S. 1998, Fiber optic device for sensing the presence of a gas, U. S. Patent No. 5,708,735.

Griessen, R., Huiberts, J. N., Kremers, M., van Gogh, A. T. M., Koeman, N. J., Dekker, J. P and Notten, P. H. L. 1997, Yttrium and lanthanum hydride films with switchable optical properties, *J. Alloys and Compounds* 253-254:44.

Lampert, C. M. 1984, Electrochromic materials and devices for energy efficient windows, *Solar Energy Mat. II* 11:1.

Crank, J. 1992, *The Mathematics of Diffusion*, Oxford Science Publishing Co., London, UK.

Hoffheins, B. S., Lauf, R. J., McKnight, T. E., Smith, R. R. and James, R. E. 1997, Evaluation of a hydrogen sensor for nuclear reactor containment monitoring, *Proceedings of the International Topical Meeting on Advanced Reactor Safety*, American Nuclear Society, 1:609.

Hoffheins, B.S., R.J. Lauf, T.E. McKnight, and Smith, R.R. 1998, Design and testing of hydrogen sensors for industrial applications, *Polymers in Sensors, Theory and Practice,* ACS Symposium Series 690, Chapter 8.

Felten, J.J. 1994. "Palladium Thick Film Conductor." U.S. Patent No. 5,338,708.

THE APPLICATION OF A HYDROGEN RISK ASSESSMENT METHOD TO VENTED SPACES

Dr. Michael R. Swain,[1] Eric S. Grilliot,[1] and Dr. Matthew N. Swain[2]

[1]University of Miami
Coral Gables, FL 33124

[2]Analytical Technologies, Inc.
Miami, FL 33186

INTRODUCTION

A comparison of the predicted results from a calibrated computational fluid dynamics (CFD) model with experimentally measured hydrogen data was made to verify the calibrated CFD model. The experimental data showed the method predicted the spatial and temporal hydrogen distribution in the garage very well. A comparison was then made of the risks incurred from a leaking hydrogen-fueled vehicle and a leaking liquefied petroleum gas (LPG)-fueled vehicle.

The following is a brief description of the use of the Hydrogen Risk Assessment Method (HRAM) to analyze the risk associated with hydrogen leakage in a residential garage. The four-step method is as follows:
1. Simulation of an accident scenario with leaking helium
2. Calibration of a CFD model, of the accident scenario, using helium data
3. Prediction of the spatial and temporal distribution of leaking hydrogen using the calibrated CFD model
4. Determination of the risk incurred by hydrogen compared to a currently used fuel.

EXPERIMENTAL VERIFICATION OF METHOD - HALLWAY

The following example is given, together with a comparison of the predicted hydrogen concentrations to the experimentally determined values.

The geometry used for this example was a half-scale hallway. The dimensions were 114 inches (2.9 m) by 29 inches (0.74 m) by 48 inches (1.22 m). Figure 1 shows a schematic of the hallway. The hydrogen escaped from the floor at one end of the hallway (left hand side of figure). A roof vent and lower door vent existed at the other end of the

Advances in Hydrogen Energy, edited by Padró and Lau
Kluwer Academic/Plenum Publishers, 2000

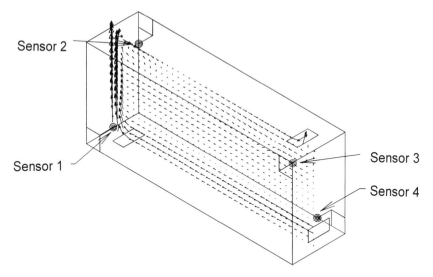

Figure 1. Hallway with velocity vectors.

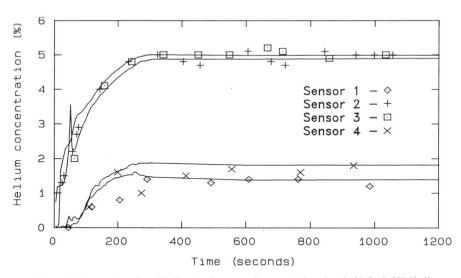

Figure 2. Comparison of model (lines) and measured concentrations (symbols) for 2 CFM heliu leak at end of hallway.

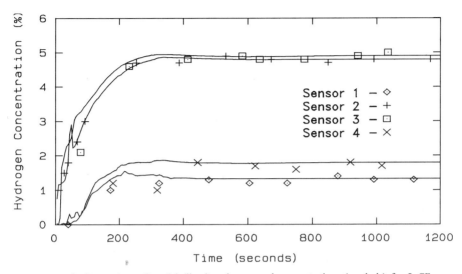

Figure 3. Comparison of model (lines) and measured concentrations (symbols) for 2 CF Hydrogen leak at end of hallway.

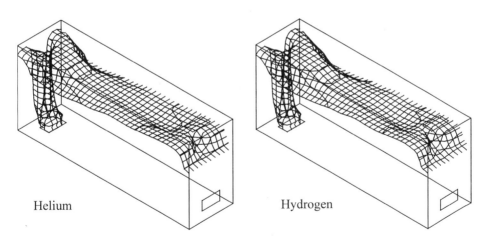

Figure 4. Helium hydrogen comparison, 2 CFM leak at end of hallway, 1 min elapsed, 1% concentration.

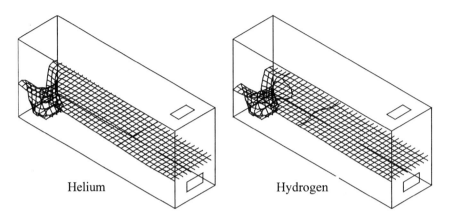

Figure 5. Helium hydrogen comparison, 2 CFM leak at end of hallway, 1 min elapsed, 1% concentration.

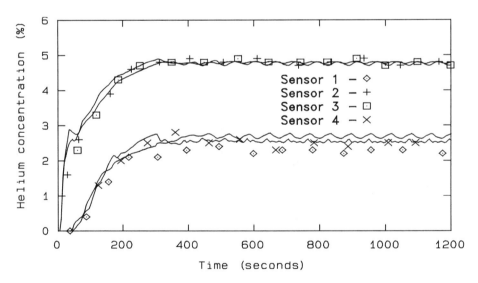

Figure 6. Comparison of model (lines) and measured concentrations (symbols) for 2 CFM Heliu leak in middle of hallway.

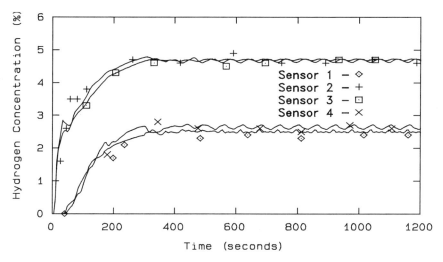

Figure 7. Comparison of model (lines) and measured concentrations (symbols) for 2 CFM Hydrogen leak in middle of hallway.

hallway (right hand side of figure). Figure 1 shows an example of the velocity vectors predicted by the CFD model. Figure 1 also shows the points at which helium or hydrogen gas concentrations were measured. Figure 2 shows the results of the CFD model compared to the experimentally measured concentrations for helium escaping at 2 SCFM from the floor at the end of the hallway. Figure 3 shows the results of the CFD model compared to the experimentally measured concentrations for hydrogen escaping at 2 SCFM from the floor at the end of the hallway. It can be seen that the CFD model predicted the hydrogen behavior accurately.

Figure 4 shows a comparison of the surfaces of constant 3% concentration for both helium and hydrogen. Both gases rise from the floor, travel across the ceiling, and leave through the roof vent. The gas leaving the roof vent causes a drop in pressure, which draws air in the lower door vent. The general circulation in the hallway can be seen in Figure 1. All of the gases inside and above the surface of constant 3% concentration contain more than 3% helium or hydrogen. Those below the surface contain less than 3% concentration.

Figure 5 shows the surfaces of constant 1% concentration for helium and hydrogen. Comparison with Figure 4 gives an indication of the vertical concentration gradient in the hallway.

Figure 6 shows the results of the CFD model compared to the experimentally measured concentrations for helium escaping at 2 SCFM from the middle of the floor in the hallway. Figure 7 shows the results of the CFD model compared to the experimentally measured concentrations for hydrogen escaping at 2 SCFM from the middle of the floor in the hallway. It can be seen that the CFD model predicted the hydrogen behavior accurately.

The method was also tested with an extended vertical vent (chimney). This was dòne to investigate a geometry that was potentially difficult to model. Figure 8 and 9 show the surfaces of constant 5% concentration versus time for both helium and hydrogen. In both cases the leakage rate was 2700 liters/hr into a 1 ft, by 1 ft, by 6 ft tall, vertical vent. The low density gas (helium or hydrogen) rises, entrains air, and forms a flow that attaches itself randomly to the four walls of the vent (see Figures 8 and 9). The concentration at a specific point was not predictable. It was therefore concluded that a vertical vent should be added to the hallway to test the ability of the CFD model to accurately predict hydrogen concentration.

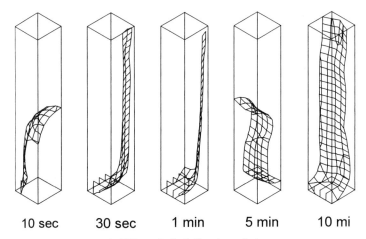

| 10 sec | 30 sec | 1 min | 5 min | 10 mi |

Figure 8. CFD results for helium in vertical vent.

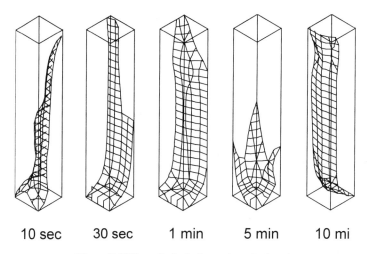

| 10 sec | 30 sec | 1 min | 5 min | 10 mi |

Figure 9. CFD results for hydrogen in vertical vent.

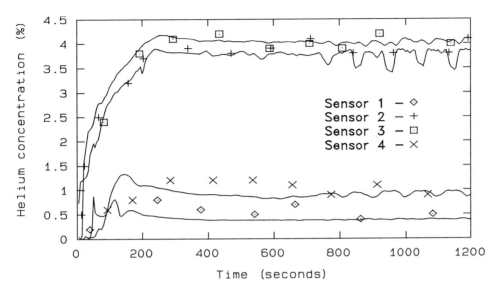

Figure 10. Comparison of model (lines) and measured concentrations (symbols) for 2 CF Helium leak at end of hallway with extended vent.

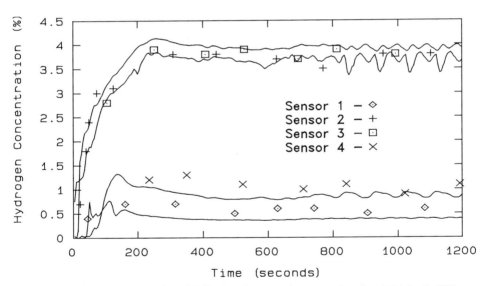

Figure 11. Comparison of model (lines) and measured concentrations (symbols) for 2 CFM Hydrogen leak at end of hallway with extended vent.

Figure 10 shows the results of the CFD model compared to the experimentally measured concentrations for helium escaping at 2 SCFM from the floor at the end of the hallway and an extended vertical vent added to the roof vent. The concentration of helium was reduced compared to the hallway without the vertical vent because the vertical vent acted as a chimney, increasing the ventilation rate in the hallway. Figure 11 shows the results of the CFD model compared to the experimentally measured concentrations for hydrogen escaping at 2 SCFM from the floor at the end of the hallway and an extended vertical vent added to the roof vent. It can be seen that the CFD model predicted the hydrogen behavior accurately.

COMPARISON OF HELIUM AND HYDROGEN CONCENTRATIONS

Though helium concentrations were slightly higher than hydrogen concentrations during this investigation, the hydrogen concentration can be greater or less than the helium concentration, depending on the enclosure geometry. Hydrogen is 8% more buoyant than helium and tends to rise more rapidly. The increased vertical velocity tends to increase both ventilation rate and gas mixing with air. If the exit vent is near the gas escape, the increased gas mixing does not decrease the concentration of hydrogen in the hydrogen-air mixture leaving through the vent enough to overcome the increased ventilation rate, and hydrogen concentration tends to be lower than helium concentration. If the exit vent is far from the escape, the increased gas mixing tends to reduce the concentration of hydrogen in the mixture leaving the enclosure enough to overcome the increased ventilation rate, and hydrogen concentration tends to be higher than helium concentration.

LPG-HYDROGEN COMPARISON (COMPUTATIONAL)

The use of a helium-verified computer model to predict gas motion can be extended to compare leakage from vehicles fueled with a variety of fuels.

The model was used to compare gas leakage from vehicles stored in residential garages. Leakage from an LPG-fueled vehicle was compared to leakage from a hydrogen-fueled vehicle. The comparison was based on a Ford Taurus sized vehicle stored in a single car garage.

Previous work for Ford Motor Company resulted in a garage door geometry that contained upper and lower, louvered and screened, vents. The garage door provided adequate ventilation to prevent accumulation of appreciable volumes of combustible hydrogen-air mixtures from a hydrogen-fueled vehicle leaking 7,200 liters/hr.

The computer model representation of the ventilated garage was run to predict the behavior of an LPG-fueled vehicle. The leakage rates chosen for the LPG-fueled vehicle were 848.2 liters/hr and 4,334 liters/hr. These represent upper and lower bounds on the leakage rate of propane from a fuel line fracture that produced a 7,200 liter/hr hydrogen leakage rate. The 848.2 liter/hr flow rate would occur if laminar flow was assumed in the hydrogen and propane leaks being compared. The 4,334 liter/hr flow rate would occur if turbulent flow was assumed in the hydrogen and propane leaks being compared. Due to differences in density and viscosity, the volumetric leakage rate of propane was lower than that of hydrogen.

Figures 12-14 show the results after 2 hours of leakage. The figures show surfaces of constant gas concentration that represents the lean limit of combustion. It can be seen that the volume of combustible gas created by the hydrogen-fueled vehicle is much smaller than the volume created by the LPG-fueled vehicle. This was true regardless of which of the two propane flow rates was assumed.

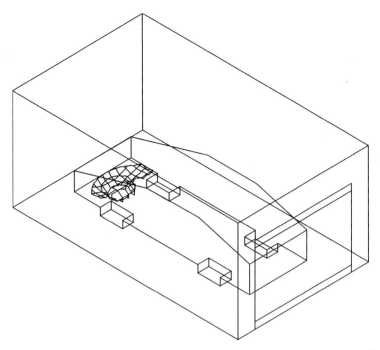

Figure 12. Surface of constant 4.1% hydrogen concentration after 2 hours of leakage at 7200 liters/hr.

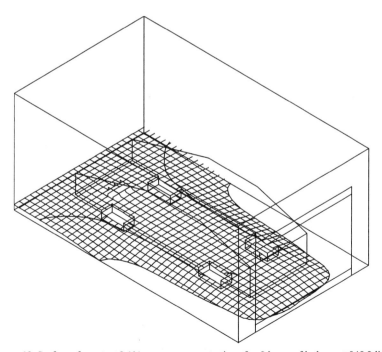

Figure 13. Surface of constant 2.1% propane concentration after 2 hours of leakage at 848.2 liters/hr.

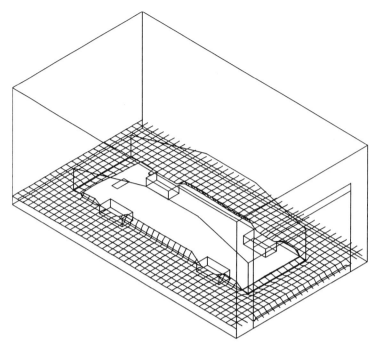

Figure 14. Surface of constant 2.1% propane concentration after 2 hours of leakage at 4334 liters/hr.

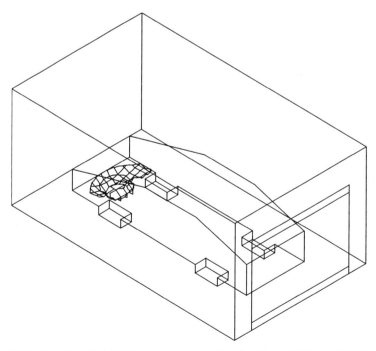

Figure 15. Surface of constant 4.1% hydrogen concentration after 1 hour of leakage at 7200 liters/hr.

Figure 12 is a plot of the surface of constant 4.1% hydrogen concentration by volume. 4.1% hydrogen in air is the upward propagating lean limit of combustion for hydrogen-air mixtures. This is the lowest concentration of hydrogen considered combustible. The cloud under the front of the vehicle in Figure 12 represents the volume of burnable gas after 2 hours of leakage at 7,200 liters/hr. The cloud had reached essentially this size at 1 hour.

Figure 13 is a plot of the surface of constant 2.1% propane concentration by volume. 2.1% propane in air is the upward propagating lean limit of combustion for propane-air mixtures. This is the lowest concentration of propane considered combustible. The cloud covering almost the entire floor of the garage represents the volume of burnable gas after 2 hours of leakage at 848 liters/hr. The cloud had reached essentially this size at 45 minutes.

Figure 14 is a plot of the surface of constant 2.1% propane concentration by volume. The cloud covering the entire floor of the garage represents the volume of burnable gas after 2 hours of leakage at 4,334 liters/hr. The cloud had reached essentially this size at 20 minutes.

The volume of combustible gases produced by the hydrogen-fueled vehicle had reached steady state after 1 hour as seen in Figure 15. Figure 15 shows the surface of constant 4.1% hydrogen concentration, which is the lean limit of combustion for hydrogen.

CONCLUSIONS

1. Accident scenarios that assume no loss of hydrogen due to buoyancy-driven convection can considerably overestimate the risk incurred in hydrogen escapes. In reality, hydrogen's very low density prevents this accident scenario from occurring.

2. A helium data verified CFD computer model can accurately predict the spatial and temporal distribution of hydrogen released in a hydrogen escape.

3. Previous work (Swain et al., 1998) showed that the risk incurred by storing a hydrogen-fueled vehicle in a residential garage was less than that of an LPG-fueled vehicle. This work has shown that the risk can be reduced further below that of the LPG-fueled vehicle, with proper garage door design.

ACKNOWLEDGEMENTS

The authors would like to acknowledge support from the Department of Energy. Without their funding, this project would not have been possible.

REFERENCES

Swain, M.R., Schiber, J., and Swain, M.N., "Comparison of Hydrogen, Natural Gas, Liquefied Petroleum Gas, and Gasoline Leakage in a Residential Garage," Energy and Fuels, Vol. 12, No. 1, 1998

MODELING OF INTEGRATED RENEWABLE HYDROGEN ENERGY SYSTEMS FOR REMOTE APPLICATIONS

Eric Martin and Nazim Muradov

Florida Solar Energy Center
1679 Clearlake Road
Cocoa, Florida 32922

INTRODUCTION

The concept of stand-alone renewable energy systems that use hydrogen as an energy storage medium has attracted much attention recently[1-3]. Such "hybrid" systems intend to utilize hydrogen technologies in order to overcome the electricity supply/demand mismatch encountered with an "off-grid" renewable power system. For example, in the case of a photovoltaic (PV) system, peak electricity supply is available in the early afternoon, while peak electricity demand for a residential application generally occurs in the morning and early evening. The main objective of this work is to develop an efficient tool to assist in the design and evaluation of an integrated solar hydrogen production/storage energy system (SHES). In general, a SHES comprises at least four major components: a PV-array, an electrolyzer, a fuel cell, and a hydrogen storage unit.

Many of the SHES demonstration projects described in the literature have involved the development or utilization of existing computer simulation software to aid in the design and modeling of the energy systems[4-6]. After analyzing the performance of several types of simulation software, and their applicability towards modeling different modifications of a SHES, the TRNSYS software package was selected as the main platform for system integration and simulation[7]. TRNSYS, created by the Solar Energy Laboratory (SEL), University of Wisconsin-Madison, is transient simulation software with source code written in FORTRAN. The transient nature of the program, where time is an inherent variable, is beneficial to this simulation for it allows the user to track system performance over the course of an entire year, with changing weather conditions and energy load profiles. The software is component driven, and total system simulation is accomplished by linking stand-alone equipment models, or components, together.

Although many of the standard TRNSYS components are useful for the SHES simulation, it should be noted that no models for the core components of the SHES (a fuel cell, an electrolyzer, and a hydrogen storage system) are currently included with the standard version of the TRNSYS software. However, the SEL has collected many applicable component models, written by various users, and has used them to create a demonstration of the capabilities of TRNSYS. This demonstration, named PVHYDRO, has been developed particularly for the design and optimization of an integrated PV-electrolyzer-fuel cell system that supplies all the domestic electricity needs of a single-family dwelling[8]. Along with routines to simulate AC/DC electricity supply and demand, the PVHYDRO demonstration also includes routines to simulate other domestic energy needs including hot water, space heating, and cooking. For hot water, a solar thermal system is included in the simulation to determine design parameters for a solar collector and hot water storage tank. PVHYDRO accomplishes space heating by a combination of methods including use of solar heated water, use of passive solar energy through windows, use of waste heat from the fuel cell, and combustion of a portion of generated hydrogen via a catalytic burner. The catalytic burner also provides energy for domestic cooking by hydrogen combustion.

This paper describes the results obtained by using TRNSYS/PVHYDRO to simulate a SHES intended to supply power to a single residence or small cluster of residences in a remote (off-grid) application. Simulations were intended to optimize the sizes of the hybrid battery/hydrogen storage components and investigate any system tradeoffs encountered. We used a FORTRAN compiler to customize the "non-standard" PVHYDRO components and link them with the main TRNSYS simulation program. Customization of the component models mainly involved refining their source code so that they accurately represented our electricity load profile and the performance of the specific equipment to be incorporated in the SHES. For the simulation of the SHES described in this paper, solar heating of water, residential heating, and catalytic combustion of hydrogen capabilities have been excluded, and we focused on AC/DC electricity supply and demand representing the remote (off-grid) application.

DESCRIPTION OF SHES

Figure 1 depicts a generalized schematic diagram of the proposed 10 kW SHES. It consists of the following main components: a PV array, an electrolyzer, a fuel cell, a battery array, a hydrogen storage unit, a controller, and an electric load. The starting point for the simulation is to determine whether the system will be able to meet the load requirements based on weather data contained in a file. The data can be supplied by the user, or TRNSYS provides typical meteorological year (TMY) data for multiple locations across the globe. These data include average values for solar insolation, ambient temperature, and humidity. For the results described in this paper, TMY data for Orlando, FL have been used.

The PV array acts as the primary producer of electricity, and it needs to be designed based on electricity demand and average values of solar insolation for the chosen location. During periods when the insolation is greater than the average, excess power generated by the array can be diverted to the electrolyzer, which effectively "stores" this excess energy by producing hydrogen, that can later be used by the fuel cell. During periods when the insolation is less than the average, the fuel cell is able to support the deficiency. In the simulation of the SHES, batteries are included for short-term storage of electricity generated by the PV array and the fuel cell. Batteries play an important role for they supply the continuous demand of electricity required by the SHES. While holding a sufficient charge, they maintain a stable and constant flow of electricity to the electric consumer regardless of system interruptions involving the PV array or fuel cell. Such interruptions include lack of proper solar insolation, lack of hydrogen supply, and power supply "gaps" encountered when transferring electricity generation between the PV array and the fuel cell.

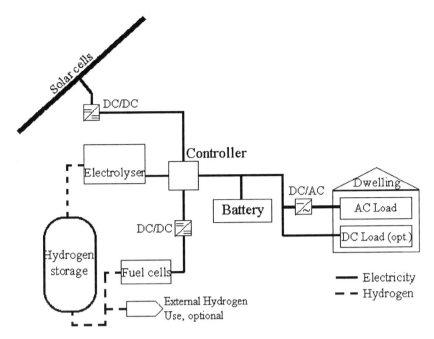

Figure 1. Schematic of SHES system.

In the interest of keeping the SHES system as renewable as possible, it is desirable to include as few batteries as possible into the system design. Batteries tend to have a much shorter lifetime (5-10 years) compared to the renewable sources of electricity found in the system (20 years). However, as the amount of battery storage capacity decreases, the parameters of other system components, such as hydrogen storage capacity and size of the PV array, increase in order to compensate. This results from lower efficiencies obtained by storing electricity produced by the PV array in the form of hydrogen (33%) compared to conventional battery storage (90%). In order to investigate this system trade-off, two simulation studies have been conducted on the proposed SHES. *Case 1* utilizes a relatively small battery array that only has the capacity to support the system's electricity needs for approximately 2 hours. This represents a situation where the batteries are only able to supply the system load when electricity production switches between the PV array and fuel cell. No means of safety back-up power is provided in the event that the PV array and fuel cell become inoperable. *Case 2* has a larger battery array, and has the capacity to support the system's electricity needs for slightly more than one day. In this case, the size of the battery array prevents the need for the fuel cell to be solely responsible for powering the load every evening, when the PV array is inactive. Instead, the fuel cell is only required to supply power when the capacity drops below a certain "safe" limit, caused by consecutive days of low solar insolation.

In a stand-alone PV system, representing a case in which a system does not utilize hydrogen as a form of energy storage, batteries are usually expected to store between 3-7 days of useable power. However, because a hybrid system such as an SHES has a secondary means of generating electricity, in this case a fuel cell, the number of days of electricity storage required from the battery array can be significantly reduced. Stand-alone systems currently benefit from having somewhat standardized design procedures and may be more cost effective than systems employing hybrid battery/hydrogen storage.

However, it is the intent of this paper to investigate systems that benefit from increased system renewability.

A single controller device oversees total system operation in the SHES simulation. By assessing the requirements and/or output available of every system component, including the electric load, the controller makes appropriate decisions to optimize system performance. These decisions include whether to connect or disconnect individual components to/from the system and whether power generated by the PV array is sent to the electrolyzer for hydrogen generation, or to the battery array for use by the electric consumer. In this respect, this component acts as a charge controller to regulate battery charging.

A generalized controller operational flowchart is included in Figure 2. This flowchart shows relative battery state-of-charge (SOC) to be the starting criteria for the decision tree. SOC is defined as the ratio of ampere-hours (Ah) stored in the battery array at any given time to the maximum Ah capacity of the array. During this process the actual controller subroutine also includes the voltage and current that is available and/or required by each component. However, for the purpose of explanation, the power available or required by various system components can be related to battery SOC.

After determining whether battery SOC is below a certain minimum value, the controller can judge whether to use available power to charge the battery, or produce hydrogen. If the battery contains a sufficient charge, and PV power is available, hydrogen can be produced by the electrolyzer and stored. If both the battery and storage tank are full at any time, the PV array will power the load directly. This may involve either partially or fully disconnecting the PV array from the load (charge control), depending on the load requirements and the amount of solar insolation available. If the battery does not contain a sufficient charge, the controller will first try to charge the battery with the PV array. If excess power from the array is available, the controller will divert the excess power to the electrolyzer such that hydrogen can be produced in conjunction with battery charging. If insufficient solar insolation is available, the controller will then use the fuel cell to power the load, through the battery array, until weather conditions permit the batteries to regain a sufficient state-of-charge via the PV array. The only instance where system downtime can occur is when the system is exclusively relying on the fuel cell for power, and the hydrogen reserve is exhausted.

RESULTS

System Components

Load Profile. The load profile used to simulate the proposed SHES reflects typical residential electricity needs in a remote (off-grid) application. In order to meet the requirements of the consumer load, the SHES must provide 58 kWh of AC electricity each day of the year, at 120 V and 60 Hz. The peak load each day is 10 kW, which occurs in the early evening, and the daily average load is 2.5 kW. The daily, peak, and average loads that the SHES system must supply also take into account losses encountered in AC/DC power inversion, DC/DC voltage conversion, and power requirements of various system components (i.e. controller).

Batteries. *Case 1* includes a total of four lead-calcium deep cycle solar batteries. The four batteries are connected in series which maintains a battery bus voltage of 48 V. The battery string has a capacity of 125 Ah, which permits the batteries to store 2 hours of required system amperage based on a daily consumption of 58 kWh.

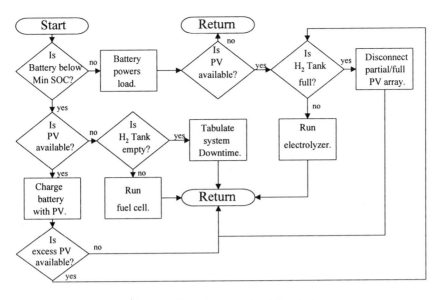

Figure 2. Controller operational flowchart.

Case 2 includes a total of 12 batteries arranged in an array of three parallel strings of four batteries each. A similar type of battery has been used in this simulation, however in this case each parallel string has a capacity of 635 Ah. With 1905 Ah total storage capacity, the battery array has the ability to solely power the load for 1.25 days, based on 58 kWh daily consumption.

Photovoltaic Array. For *Case 1*, a total of 588 individual modules are used in the simulation, and the array is configured with 147 parallel strings of 4 modules each. The panels are tilted at an angle of 45° to optimize electricity production during the winter months. Using the manufacturer's guaranteed power output of 100 W per module at standard test conditions (irradiance = 1000 W/m^2, cell temperature = 25°C), the peak power output of the array is expected to be 58.8 kW.

For *Case 2*, a total of 256 of the same modules are used, and the array is configured with 64 parallel strings of 4 modules each. The panels are also tilted at an angle of 45°, and the peak output of the array at standard test conditions is expected to be 25.6 kW.

Electrolyzer. Data for a PEM type pressurized electrolyzer has been used for the SHES simulation. The same electrolyzer has been used for both cases with a total of 25 cells and each cell is assumed to have an area of 279 cm^2. The electrolyzer operates at a pressure of 1000 psi and an efficiency of approximately 75%, producing approximately 0.48 kg/h (90 slpm) of hydrogen at 500 A.

Hydrogen Storage. The SHES simulation includes a pressurized tank that stores hydrogen as it is produced by the electrolyzer. The maximum pressure of the tank is 1000 psi. For *Case 1*, an optimized storage volume of 20 Nm^3 is used in the SHES simulation, allowing a maximum of 90 kg of hydrogen to be stored at ambient (25°C) temperatures. For *Case 2*, an optimized storage volume of 10 Nm^3 is used, allowing a maximum of 45 kg of hydrogen to be stored at 25°C.

Fuel Cell. A PEM fuel cell is included in the SHES simulation for secondary power generation. The same fuel cell has been used for both cases. It operates on hydrogen and air and contains a total of 50 cells. Each cell has an area of 300 cm^2. The stack produces a total of 11.6 kW of DC power at 32 V and 363 A. The stack operates at an efficiency of 44% consuming 0.683 kg/h (127 slpm) of hydrogen.

Power Conditioning. In the SHES simulation, four power conditioning devices are included (see Figure 1). A maximum power point tracker (MPPT) maintains optimum performance of the PV panels by ensuring that the array operates at the maximum power point on its I-V curve. A DC to DC converter upgrades the fuel cell output voltage to the battery bus voltage. A diode prevents the back flow of current from the battery array and fuel cell to the electrolyzer. This ensures that the only source of power for the electrolyzer is the PV array. Finally, a DC to AC inverter is included to invert the DC power supplied by the battery array to AC power required by the electric consumer. The efficiency of all power conditioning devices is assumed to be approximately 90%.

Table 1. Summary of optimized component parameters determined from simulations

Component Parameter	Case 1	Case 2
Battery Array		
Number of batteries	4	12
Array capacity	125 Ah	1905 Ah
Days of storage	0.083 d (2 hr)	1.25 d
Hydrogen Storage		
Tank volume	20 m^3	10 m^3
PV Array		
Number of modules	588	256
Array power (Pmax)	58.8 kW	25.6 kW
Fuel Cell		
Type	PEM	PEM
Efficiency	44%	44%
Power	11.6 kW	11.6 kW
Electrolyzer		
Type	PEM	PEM
Efficiency	75%	75%

System Performance

Figures 3 and 4 depict the system performance of *Case 1* for the first weeks of February and August respectively. In these plots the power produced by the PV array is shown along with the power requirement of the system load and battery state-of-charge (SOC). Figures 5 and 6 show system performance for the first weeks of February and August for *Case 2*, respectively.

Examples of controller operation are shown in Figures 7 and 8 for the *Case 1* and Figures 9 and 10 for *Case 2*, for the first weeks of February and August, respectively. These figures show how the controller decides when to operate the electrolyzer and fuel cell, based on the battery SOC. The plots for the electrolyzer show the power used by this component, which has been drawn from the PV array, and the plots for the fuel cell show the power produced by this component. Instances when the fuel cell is not operating at full capacity can be seen in these plots, and they are a result of how the controller's subroutine was originally written. The original PVHYDRO demonstration system contains water and space heating requirements, along with the electric load, and the system relies on the fuel cell to generate a portion of this heat during operation. In order to maximize fuel cell heat production, when the battery contains a sufficient state-of-charge (i.e. the fuel cell is not needed for power generation) but the heating load is large, a small fraction of the total number of individual fuel cells are activated such that heat can be generated.

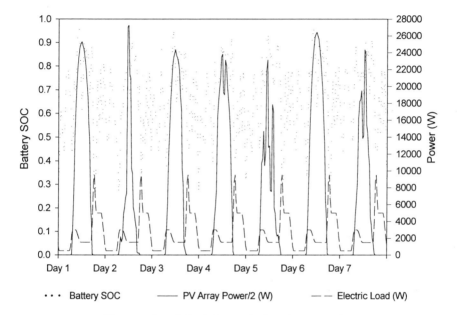

Figure 3. Case 1 simulation results for February, week 1.

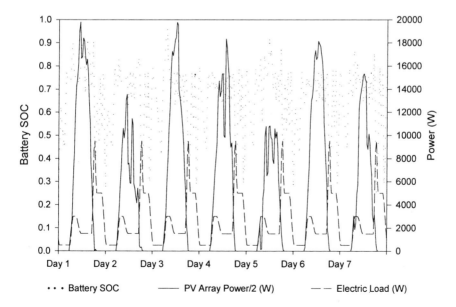

Figure 4. Case 1 simulation results for August, week 1.

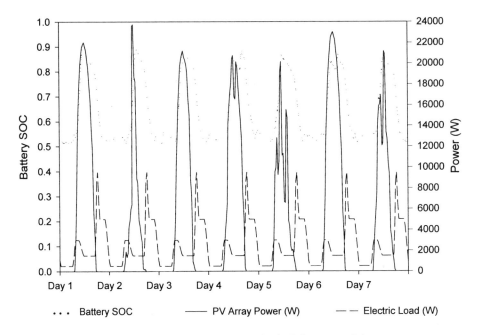

Figure 5. Case 2 simulation results for February, week 1.

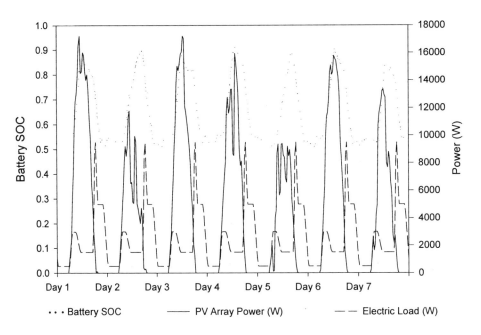

Figure 6. Case 2 simulation results for August, week 1.

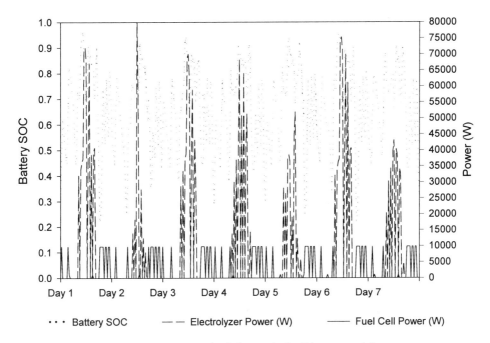

Figure 7. Case 1 simulation results for February, week 1.

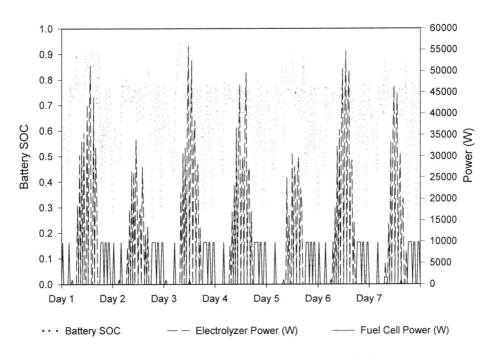

Figure 8. Case 1 simulation results for August, week 1.

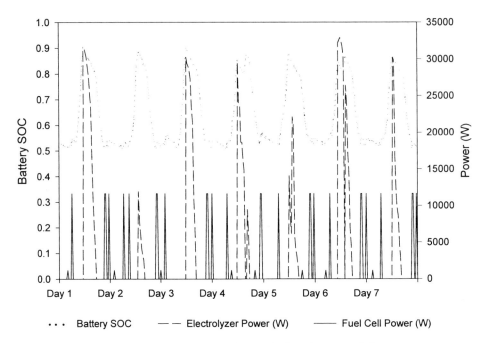

Figure 9. Case 2 simulation results for February, week 1.

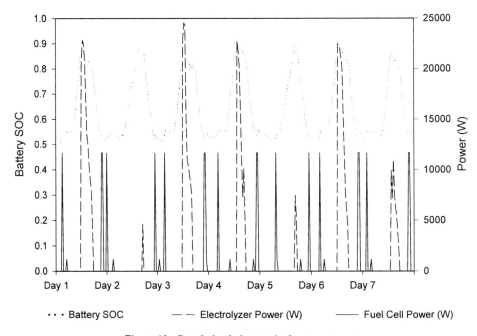

Figure 10. Case 2 simulation results for August, week 1.

Figures 11 and 12 show examples of how the system produces, stores, and uses hydrogen for *Case 1* during the first weeks of February and August, respectively. Figures 13 and 14 show similar data for *Case 2*. Each of these plots shows the power drawn by the electrolyzer, and how the amount of hydrogen produced affects the value for hydrogen storage. The plots also show the power produced by the fuel cell, and how the amount of hydrogen consumed affects the amount of hydrogen reserve. As with battery SOC, the parameter for hydrogen storage is represented as the ratio of the amount of hydrogen in the tank at any time to the maximum capacity of the tank.

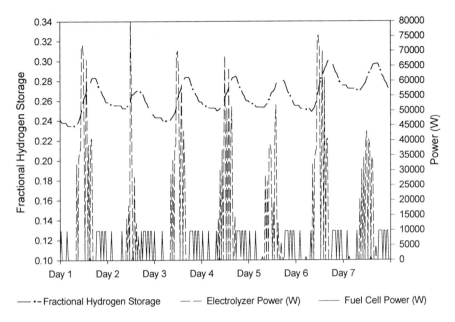

Figure 11. Case 1 simulation results for February, week 1.

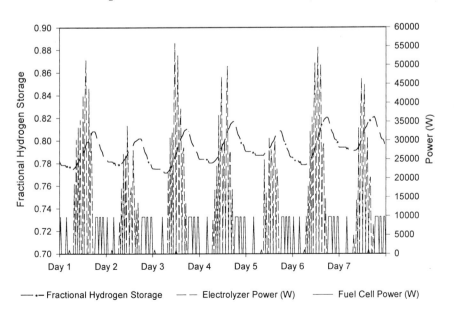

Figure 12. Case 1 simulation results for August, week 1.

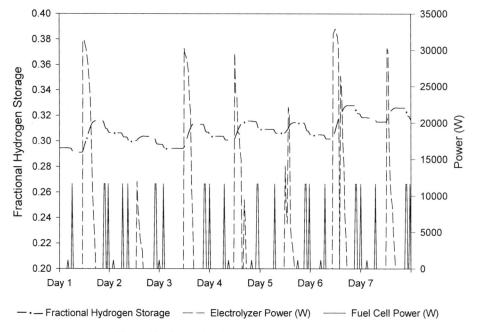

Figure 13. Case 2 simulation results for February, week 1.

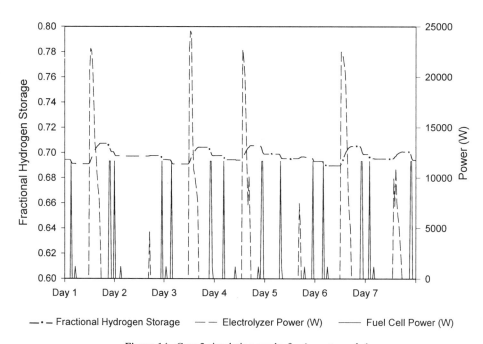

Figure 14. Case 2 simulation results for August, week 1.

Figure 15 shows simulation results for hydrogen storage over the course of an entire year for *Case 1*, starting at the beginning of January and terminating at the end of December. Figure 16 shows these results for *Case 2*. As can be seen in these plots, values for hydrogen storage fluctuate seasonally with a minimum value occurring in February and a maximum value occurring in August.

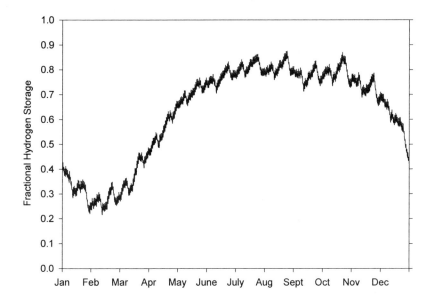

Figure 15. Case 1 simulation results: monthly variation in hydrogen storage.

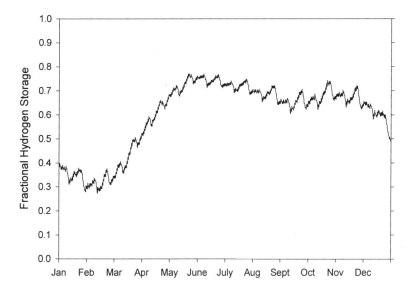

Figure 16. Case 2 simulation results: monthly variation in hydrogen storage.

DISCUSSION

Along with the design of each individual SHES component, total system operation plays a major role in system performance, size, and cost. Each of the individual components are intimately linked together in some way, therefore changing the size or operational parameters of one component has the potential to disturb the balance of the entire system. Simulations of the proposed SHES have been conducted with the intent of optimizing system performance, rather than cost, however it is expected that an optimized system will also be the most cost effective. The primary parameter that is intended to be the basis for optimization is the value for the volume of the hydrogen storage tank. For these simulations, the system was considered optimized if the value for hydrogen storage was the same at both the beginning and end of the one-year simulation. Other mandatory system criteria included use of a PEM type fuel cell and electrolyzer, maximizing battery lifetime, and maintaining zero percent system downtime.

Depending on what time of year the start-up of the SHES system occurs, the hydrogen tank should be initially charged with the amount of hydrogen found in Figures 15 or 16 that corresponds to that particular time of year. The optimized storage tank values of maximum pressure and storage volume ensure that the tank is never completely empty, and rarely completely full to account for expected variation in weather conditions. Starting up the SHES with a different value than that found in Figures 15 or 16 would perturb this balance. Starting with more hydrogen than necessary would not cause system downtime, but the system would be over-designed and would not utilize the full potential of each system component. Starting with less hydrogen than depicted would eventually cause system downtime during the winter season.

The criteria used by the controller to decide when to operate the fuel cell and electrolyzer has also been optimized. Running the electrolyzer slightly before the battery array has reached its maximum state-of-charge effectively lowers the PV power requirement, thereby reducing the size of the array. Controller settings for how often the fuel cell operates are also important operational parameters. Since the fuel cell represents mainly a capital expense, it is desirable to use it to generate power as often as possible, rather than just during the winter. Therefore the PV array is tilted at an angle of $45°$ which optimizes winter time performance of the panels, when there is less solar insolation available, according to the sun's winter time position in the sky. This arrangement requires use of the fuel cell also during the summer months, rather than leaving an expensive piece of equipment to lie dormant for several months. The controller can also be programmed to operate the fuel cell more often than in the current simulations to put less of a demand on the PV array for power production. However, in this situation, the size of the array would actually increase to supply the increased hydrogen demand, and the size of the storage tank would also increase.

As seen by the results of simulations for *Cases 1* and *2*, a system trade-off exists among battery array storage capacity, hydrogen tank storage capacity, and the size of the PV array. With a smaller battery array, less energy can be stored in this short-term medium, and therefore the size of the hydrogen tank must increase in order to store enough energy for use when the PV array is inactive. Because hydrogen production is initiated with the electricity produced by the PV array, the size of the PV array must increase in order to support daytime electricity production as well as the increased hydrogen demand.

The differences in fluctuations of battery SOC between *Case 1* and *Case 2* can be seen in Figures 3 and 4, and Figures 5 and 6, respectively. The daily trends shown in these plots remain relatively constant over the course of the entire year. As seen from the results for *Case 2*, the battery array is able to maintain a maximum depth of discharge of approximately 50%. Incorporating a battery array large enough such that excessive discharge depths can be avoided will prolong the life of the battery array, and provide some back-up power (i.e. the unused 50% of the array storage capacity) in the event of PV array/fuel cell interruptions. However, as seen from the results for *Case 1*, the smaller battery array is required to discharge to much deeper levels, reducing its lifetime and the amount of back up power available.

For future work, improvements can be made to the controller subroutine such that simulation results for systems that employ small battery arrays can be more easily interpreted. With the current controller subroutine, the smallest practical simulation step size is 0.5 hours. However, with a small battery array, it is possible that the battery SOC can change dramatically within that time. Since the controller currently bases most decisions on the SOC value, fuel cell and electrolyzer activation in such a simulation appear to be unsteady and erratic. Examples of this behavior are seen in *Case 1* results where wide variations in battery SOC cause the electrolyzer and fuel cell to switch on/off repeatedly, rather than maintaining steady operation.

Since the intent of this simulation work was to optimize system performance, a number of important issues related to the energy storage sub-system were left until the cost optimization of the SHES. Among them are the optional use of metal hydrides as a hydrogen storage medium (instead of pressurized hydrogen) and storage of oxygen produced by the electrolyzer for the use in the fuel cell. Since the same amount of hydrogen is required regardless of the type of storage medium, the use of a different medium may affect system cost, but not performance. Utilizing pure oxygen in the fuel cell rather than ambient air will decrease the size, and therefore cost of the fuel cell component. However, the cost of adding an oxygen storage component may surpass the savings encountered from reducing the size of the fuel cell.

CONCLUSION

TRNSYS was chosen as a viable platform for performing simulations on the proposed SHES. The main components of the energy system are a photovoltaic array, PEM electrolyzer, PEM fuel cell, battery, pressurized hydrogen storage unit, controller, and electric load. The simulation code was customized in order to model the specific characteristics of proposed system components. A realistic load profile, with a daily peak of 10 kW and daily average of 2.5 kW, was chosen as an example application for the renewable energy produced by the SHES system, and system components have been designed and optimized to meet this load with zero percent system downtime. Results from simulations of two cases, one with a battery array storage capacity of 125 Ah and one with 1905 Ah, demonstrate that a system trade-off exists among size of the PV array, hydrogen storage capacity, and battery array storage capacity. A system with a smaller battery array storage capacity requires a larger PV array and hydrogen storage tank to support the system's electricity requirements, and may be more costly. The details of this cost analysis, along with the potential to store oxygen produced by the electrolyzer and the use of metal hydrides to store hydrogen, have been left until a system cost optimization is conducted.

ACKNOWLEDGEMENTS

The authors would like to acknowledge the technical support provided by the Solar Energy Laboratory/University of Wisconsin and the financial support provided by U.S. Department of Energy, Energy Partners, L.C., Caribbean Marine Research Center, and the Florida Solar Energy Center™.

REFERENCES

1. P.A. Lehman and C.E. Chamberlin, Design of a photovoltaic-hydrogen-fuel cell energy system, in: *International Journal of Hydrogen Energy*. 16 (5): 349-352 (1991).
2. S. Galli, G. De Paoli, and A. Ciancia, An experimental solar PV-hydrogen-fuel cell energy system: preliminary results, in: *Hydrogen Energy Progress X*, D.L. Block and T.N. Veziroglu, eds., Proceedings of the tenth World Hydrogen Energy Conference, Cocoa Beach, FL (1994).
3. H. Barthels, W.A. Brocke, K. Bonhoff, et al., Phoebus-Julich: an autonomous energy supply system comprising photovoltaics, electrolytic hydrogen, fuel cell, in: *Intl. J. of Hydrogen Energy*. 23 (4): 295-301 (1998).
4. C.E. Gregoire Padro, V.L. Putsche, and M.J. Fairlie, Modeling of hydrogen energy systems for remote applications, in: *Hydrogen Energy Progress XII*, J.C. Bolcich and T.N. Veziroglu, eds., Proceedings of the twelfth World Hydrogen Energy Conference, Buenos Aires, Argentina (1998).
5. J. Vanhanen and P. Lund, Computational approaches for improving seasonal storage systems based on hydrogen technologies, in: *International Journal of Hydrogen Energy*. 20 (7): 575- 85 (1995).
6. S.O. Morner, W.A. Beckman, and S.A. Klein, Comparison of simulations of a photovoltaic- hydrogen system with measurements from the Schatz Solar Hydrogen project, in: *Hydrogen Energy Progress X*, D.L. Block and T.N. Veziroglu, eds., Proceedings of the tenth World Hydrogen Energy Conference, Cocoa Beach, FL (1994).
7. S.A. Klein, W.A. Beckman, et al., TRNSYS: A transient system simulation program. Solar Energy Laboratory, University of Wisconsin-Madison, WI 53706 (1992).
8. PVHYDRO demonstration website: http://sel.me.wisc.edu/trnsys/demo/pvhydro.htm.

INDEX